IoT Signal Detection

IEEE Press
445 Hoes Lane
Piscataway, NJ 08854

IoT Signal Detection

Rui Han
Beihang University

Jingjing Wang
Beihang University

Lin Bai
Beihang University

Jianwei Liu
Beihang University

IEEE Press
WILEY

Published by John Wiley & Sons, Inc., Hoboken, New Jersey.
Published simultaneously in Canada.

Library of Congress Cataloging-in-Publication Data applied for:
Hardback ISBN: 9781394183081

Cover Design: Wiley
Cover Image: © Yuichiro Chino/Getty Images

Set in 9.5/12.5pt STIXTwoText by Straive, Chennai, India

To our families and friends

Contents

List of Figures

List of Algorithms

About the Authors

Rui Han received the PhD degree in cyber security from Beihang University, Beijing, China, in 2022. From 2022 to 2024, Dr. Han was a research fellow at National Research Center, Tsinghua University, Beijing, China. Her current research interests include the Internet of things (IoT), unmanned aerial vehicle (UAV) communications, and satellite communications.

Jingjing Wang received his BSc degree in electronic information engineering from the Dalian University of Technology, Liaoning, China, in 2014 and the PhD degree in information and communication engineering from the Tsinghua University, Beijing, China, in 2019, both with the highest honors. From 2017 to 2018, he visited the next-generation wireless group chaired by Prof. Lajos Hanzo in the University of Southampton, UK. Dr. Wang is currently a professor at the School of Cyber Science and Technology, Beihang University, Beijing, China. His research interests include AI-enhanced next-generation wireless networks, UAV networking, and swarm intelligence. He has published over 100 IEEE Journal/Conference papers. Dr. Wang was a recipient of the Best Journal Paper Award of IEEE ComSoc Technical Committee on Green Communications & Computing in 2018, and the Best Paper Award of IEEE ICC and IWCMC in 2019. He is currently serving as an editor for the *IEEE Wireless Communications Letter* and the *IEEE Open Journal of the Communications Society*. He has served as a guest editor for *IEEE Internet of Things Journal*.

Lin Bai received the BSc degree in electronic and information engineering from the Huazhong University of Science and Technology, Wuhan, China, in 2004, the MSc degree (Hons.) in communication systems from the University of Wales, Swansea, UK, in 2007, and the PhD degree in advanced telecommunications from the School of Engineering, Swansea University, UK, in 2010. Since 2011, he has been with Beihang University (Beijing University of Aeronautics and Astronautics, BUAA), Beijing, China, where he is currently a professor at the School of Cyber Science and

Technology. His research interests include the security of space-air-ground integrated network (SAGIN), broadband wireless ad hoc network, unmanned aerial vehicle (UAV) communications, and Internet of Things (IoT). He has authored two books published by Springer in 2012 and 2014. He was the Symposium Co-Chair of IEEE GLOBECOM 2019, IEEE VTC 2021, and IEEE/CIC ICCC 2024, the Tutorial Co-Chair of IEEE/CIC ICCC 2019. He is the founding chair of IEEE ComSoc Wireless Communications Technical Committee Special Interest Group (SIG) on Space Air Ground Integrated (SAGI) Communications. He has served as an editor for *IEEE Transactions on Signal Processing* and *IEEE Wireless Communications Letters*, a lead guest editor for *IEEE Wireless Communications*, and a guest editor for IEEE Internet of Things Journal. He is currently serving as an editor for *IEEE Transactions on Wireless Communications* and *IEEE Transactions on Mobile Computing*, and the managing editor for *Journal of Communications and Information Networks*. He is a distinguished lecturer of the *IEEE Communications Society* and the *IEEE Vehicular Technology Society*.

Jianwei Liu received the BSc and MSc degrees in electronic and information from Shandong University, Shandong, China, in 1985 and 1988. He received the PhD degree in communication and electronic system from Xidian University, Shaanxi, China, in 1998. Currently, he is a professor with the School of Cyber Science and Technology, Beihang University, Beijing, China. His research interests include wireless communication network, cryptography, and network security.

Foreword

Over the past decades, Internet of Things (IoT) has been well developed to become one of the most important technologies in the 21st century, which aims to provide heterogeneous services for massively connected devices. Evidently, massive connectivity in IoT causes severe access congestion, and signal collision and signal superposition occur frequently. Therefore, signal detection becomes crucial in IoT communication systems. This book provides a range of key techniques to support massive IoT devices, and various signal detection techniques are explained in the context in terms of sparse signal detection, collided signal detection, round-trip delay estimation, backscatter signal division, etc. It makes an easy-to-follow presentation from the elementary to the profound level with a carefully balanced blend of theoretical elements and applications.

My colleagues, Dr. Han, Prof. Wang, Prof. Bai, and Prof. Liu, have worked on this topic for many years. They have made good achievements and published a number of papers within this topic. This book provides fundamentals of signal detection and estimation together with new results that have been developed for IoT applications, which is ideal for both graduate students and practicing engineers in wireless communications.

Academician
Chinese Academy of Engineering
Beijing

Quan Yu

Preface

Machine-type communication (MTC) is expected to play a crucial role in supporting a number of devices for Internet of Things (IoT). Due to the fact that most IoT devices have sparse activity and low signaling overhead, random access (RA) can be employed for MTC to provide an efficient way to support massive IoT devices with minimized network overload.

However, there exist many problems in IoT RA, e.g., signal collision, signal superposition. In order to face these challenges, we focus on the signal detection for IoT in terms of sparse signal detection, collided signal detection, round-trip delay estimation, and backscatter signal division.

Our book mainly focuses on the signal detection for RA in IoT, which covers the fundamentals of signal detection with two chapters dedicated to important background materials. Besides, various signal detection and estimation techniques are explained, e.g., variational inference algorithm, compressive sensing reconstruction algorithm, and we include a number of recent research outcomes that are useful for those experts in this area. In addition, the techniques are then analyzed using performance analysis tools, and simulation results are also given to help readers to understand the theorem and algorithm.

So far, there are many existing books related to signal detection. To be different from those books, our book focuses on signal detection for RA in IoT systems. Although our book is very specific, we have adopted an easy-to-follow presentation from the elementary to the profound level. We include a number of recent research outcomes that are useful for those experts in this area, where the methods of signal detection have low complexity and are suitable for IoT massive access scenarios. Meanwhile, signal processing approaches to detect the event of collision or to estimate the arrival time of multiple signals are also considered in this book. Besides, our book includes more illustrations of the signal detection results to help readers to understand the theorem and algorithm.

Our group has worked on the design of signal detection for many years and has produced various new research outcomes on IoT signal detection. In this book, the fundamentals of signal detection are covered, which is ideal for both graduate students and practicing engineers in wireless communications. In addition, this book includes not only our research outcomes but also other recent research outcomes that could be very useful to practitioners and postgraduate students who want to learn new outcomes of IoT signal detection in the field of signal processing.

Beijing
October 2024

Rui Han
Jingjing Wang
Lin Bai
Jianwei Liu

Acknowledgements

We would like to thank many people for supporting this work, in particular: Q. Yu (Chinese Academy of Engineering), J. Choi (Deakin University), W. Zhang (University of New South Wales). They helped us by providing valuable comments and useful discussions.

Special thanks go to those people who inspire and encourage us all the time: J. Lu (Tsinghua University) for guidance and encouragement, C. Jiang (Tsinghua University) for generous support, J. Wang for long-term friendship, and many others including our students, J. Ma, Q. Zeng, Y. Shan, Y. Yu, M. Liu, C. Liu, Y. Li, Y. Wang, for careful proofreading.

Then, we want to express our appreciation to our parents, families, and friends. Without their support, we can barely make the achievement.

Finally, we deeply thank Editor B. Kurzman and S. Indirakumari, who were always there with us, for their wonderful help during the completion of the book.

Acronyms

ACB	access class barring
AMC	automatic modulation classification
AMP	approximate message passing
AWGN	additive white Gaussian noise
BER	bit error rate
BP	basis pursuit
BPSK	binary phase shift keying
CAVI	coordinate ascent variational inference
CB	conjugate beamforming
CDMA	code division multiple access
CLT	Central Limit Theorem
CS	compressive sensing
CSCG	circular symmetric complex Gaussian
CSI	channel state information
DA	device association
EAB	extended access barring
FBMC	filter bank multicarrier
FD	full-duplex
FJT	fully-coordinated joint transmission
F-OFDM	filter-based orthogonal frequency division multiplexing
FPC	fractional power control
GA	Genetic Algorithm
GLRT	generalized likelihood ratio test
HD	half-duplex
HOC	high-order cumulant
IA	interference alignment
ISI	inter-symbol interference
LASSO	least absolute shrinkage and selection operator
LDPC	low-density parity check

MAP	maximum a posteriori
MCMC	Markov chain Monte Carlo
MIMO	multiple-input multiple-output
ML	maximum likelihood
MMSE	minimum mean square error
NFV	network function virtualization
NOMA	non-orthogonal multiple access
OFDM	orthogonal frequency division multiplexing
OMP	orthogonal matching pursuit
OOK	on–off keying
PA	power amplifier
PBF	partially-coordinated beamforming
pdf	probability density function
PEP	pairwise error probability
QAM	quadrature amplitude modulation
QPSK	quadrature phase shift keying
SER	symbol error rate
SIC	successive interference cancellation
SINR	signal-to-interference-plus-noise ratio
SNR	signal-to-noise ratio
SM	spatial modulation
SORTE	second-order statistic of the eigenvalues
VI	variational inference
ZF	zero forcing
ZFB	zero-forcing beamforming
4G	fourth generation
5G	fifth generation

1

Introduction

- In the present era, as Internet applications continue to evolve, studies on Internet of Things (IoT) have represented a growing field. This is particularly driven by the advancements in the latest generation of information technology, which leads to various innovative paradigms such as smart agriculture, smart health, and smart logistics. As a result, the concept of intelligent interconnectivity between all objects has become a reality. This has led to a profound transformation in the way people live, work, and travel, ushering in a new era of possibilities.

1.1 IoT in 5G

1.1.1 What Is IoT

IoT refers to the technique of using various devices, such as radio frequency identification (RFID), sensors to collect necessary information of things or progresses, then transmitting the collected data through the network, to achieve the ubiquitous connection between objects and people, objects and objects [1].

The concept of IoT was initially proposed by the Massachusetts Institute of Technology in 1999 [2]. At its early stages, IoT referred to the network created by the RFID technology and equipment. By incorporating the Internet and adhering to agreed-upon communication protocols, this network facilitated intelligent identification and management of product information, enabling seamless interconnectivity [3].

However, with ongoing technological advancements and application development, the scope of IoT has been broadened. In its modern aspect, IoT encompasses the integration of perception, identification, and control of interconnected objects. This integration, combined with networking and intelligent processing capabilities, enables the formation of highly intelligent decision-making systems [4].

IoT Signal Detection, First Edition. Rui Han, Jingjing Wang, Lin Bai, and Jianwei Liu.

As outlined in the White Paper on IoT published by the China Academy of Information and Communications Technology (CAICT) [5], IoT represents an expansive application and network extension of existing communication networks and the Internet. Through the utilization of perceptual technology and intelligent devices, IoT enables the perception and identification of the physical world. By leveraging network transmission and interconnectivity, IoT facilitates computation, processing, and knowledge mining. As a result, it enables seamless interaction and connectivity between individuals and objects, as well as between objects themselves. This, therefore, enables real-time control, precise management, and informed decision-making processes.

Regarding the fifth generation (5G) of wireless technology, advancements of IoT primarily stem from innovations in wireless and network technologies [6]. Within the field of wireless technology, the industry has placed particular emphasis on large-scale antenna arrays, ultra-dense networking, innovative multiple access techniques, and full spectrum access. Network technology has also witnessed significant progress, with the widespread recognition of a new network architecture based on software-defined networking (SDN) and network function virtualization (NFV) [7].

Furthermore, several key technologies have been exploited as important and promising contributors to 5G, including filter-based orthogonal frequency division multiplexing (F-OFDM), filter bank multicarrier (FBMC), full duplex and flexible duplex, multivariate low-density parity check (LDPC) codes, network coding, and polarization codes.

From a network architecture perspective, 5G inherits the overall characteristics of the fourth generation (4G), encompassing access networks, core networks, and upper-layer applications. However, to accommodate the diverse requirements of IoT, 5G has introduced new key technologies in both the core network and access network domains, bringing on technological innovations and network advancements.

1.1.2 Applications of IoT

Currently, there has been a successful combination between 5G communication technology and IoT [8]. This integration has brought on widespread applications in various domains including intelligent manufacturing, wireless healthcare, and smart logistics. The inherent advantages of 5G and IoT, such as enhanced efficiency and reliability, have significantly propelled the improvement of practical applications in these areas. The utilization of 5G-based IoT is expected to incredibly enhance the overall quality of life and transform our work and lifestyle practices fundamentally.

In recent years, the widespread adoption of IoT, particularly in the industrial area, has become increasingly prevalent [9]. A prime example of the combination between 5G communication technology and IoT is the application of wireless healthcare, which has greatly facilitated the life of individuals. The integration of IoT technology into medical devices has brought on an escalating demand for superior communication quality in remote control scenarios [10].

During the era of 4G communications, exposure to issues of signal quality has been shown to be related to adverse effects on the continuity and safety of the entire diagnostic and treatment process. Thankfully, the advent of 5G networks has significantly improved signal quality, offering low-latency and reliable connections. This enhancement enables medical professionals to accurately and promptly assess patients' conditions in remote diagnosis, thanks to high-resolution images and videos, and provide precise feedback in wireless healthcare. Besides, the utilization of remote medical devices in scientific surgical procedures ensures a high-quality service for patients.

Moreover, 5G communications have emerged as powerful platforms for autonomous driving systems. With the benefits of 5G, autonomous vehicles can strengthen their perception, decision-making, and control functions to the edge cloud. This architecture allows the processing and dissemination of data to be carried out utilizing the computational capabilities of the edge cloud, thus reducing the reliance on vehicle sensors [11].

Furthermore, the remote driving within the field of Internet of Vehicles (IoV) becomes a key instrument in efficient vehicle control and monitoring from production scheduling centers. The incorporation of remote driving not only ensures the driving safety and well-being of individuals but also gives significant advancements within the automotive industry, which promotes convenient transportation for individuals and generates economic benefits for the automotive sector.

1.1.3 Future of IoT

The Narrowband Internet of Things (NB-IoT) technique has been a major contributor in driving the widespread implementation of IoT within various domains, e.g., industrial interconnection, and improving people's quality of life. This advancement has led to the establishment of a innovative industrial ecosystem and substantial commercial expansion [12].

Looking ahead, the future of IoT promises to experience significant transformations in the next two decades due to the continuous development of innovative technologies. One prominent example is the forthcoming realization of large-scale vehicle and utility automation. This progress will encompass various facets, such as smart grids for energy production, efficient waste management systems, and intelligent environmental monitoring, all aimed at reducing greenhouse gas emissions and pollution.

While the utilization of 5G network in IoT offers numerous benefits and shows remarkable potential for future advancements, it is crucial to acknowledge the issues that arise during the practical implementation of a 5G-based IoT architecture. These issues include inadequate communication security, significant investment costs, increasing energy consumption, and limited availability of high-frequency resources, all of which contribute to the complexities associated with deploying IoT networks.

These difficulties primarily stem from the expansion of network infrastructure, effective management of IoT equipment, and efficient processing of the vast amount of data generated by the extensive deployment of mechanical devices within the IoT ecosystem. Furthermore, interoperability and heterogeneity pose additional challenges within the context of IoT. For instance, seamless integration of heterogeneous networks remains a persistent issue, inhibiting effective connectivity among these networks. In particular, the interoperability problem is increasingly recognized as a serious concern in the communication and exchange of information between numerous mechanical devices and intelligent networks, as well as in establishing connections with various applications. The lack of standardized protocols and compatible interfaces hinders smooth interoperability. Moreover, there are potential concerns regarding information security. Ensuring data integrity and confidentiality within the IoT communication system presents a significant challenge, as vulnerability in the network infrastructure and device ecosystem may compromise the security of the transmitted data.

To effectively tackle the challenges arising from the application of IoT technology in the 5G network and ensure the efficient utilization of IoT systems, it is imperative to establish unified standards for wireless communications, data collection, and security management. These standards should satisfy the specific requirements of 5G IoT applications and facilitate the advancement of robust security management and control mechanisms.

Currently, researches on 5G-based IoT networks are in their early stages and lack a mature research framework and content. Given this, it is crucial to dedicate efforts towards exploring and developing practical technologies for 5G-based IoT systems to satisfy the growing demand of IoT communications. Simultaneously, addressing existing challenges requires enhancing the heterogeneity and mobility of 5G network applications to cater to the scale and characteristics of IoT devices. By doing so, we can effectively tackle the existing obstacles.

1.2 IoT Networks

The IoT network architecture consists of the sensor/perception layer, the network layer, the communication layer, the architecture layer, and the application layer, as shown in Fig. 1.1, which are introduced as follows:

- The sensor/perception layer realizes intelligent perception, information collection and processing, automatic control of the physical world, and connects the

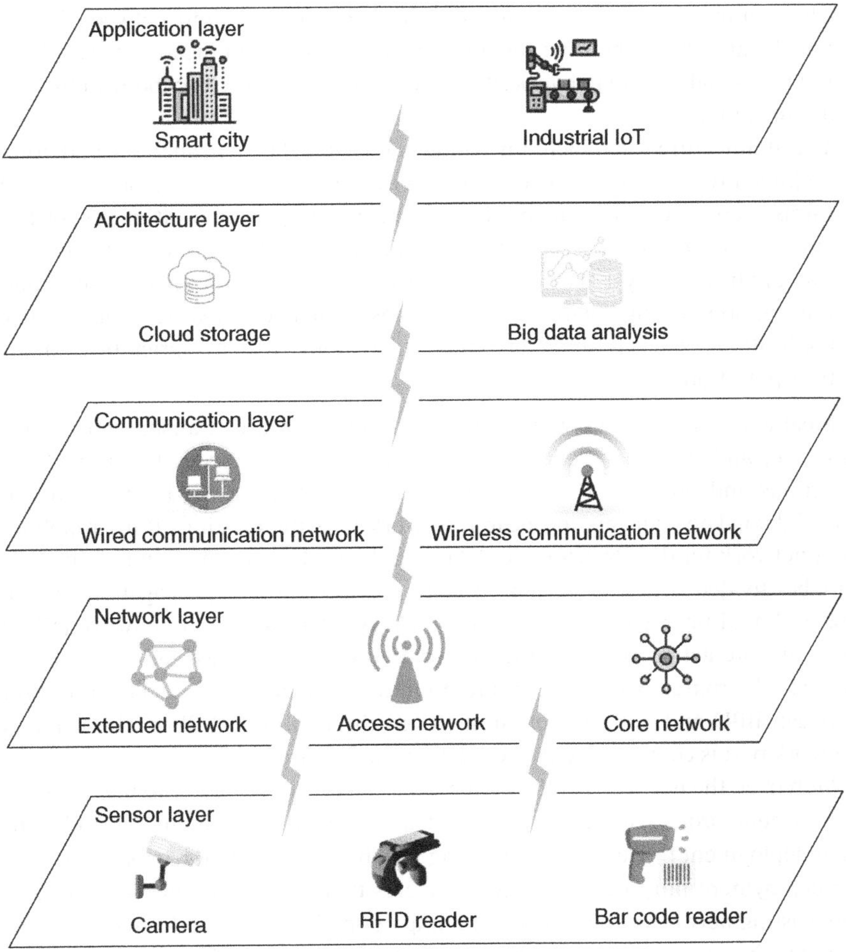

Figure 1.1 Architecture of IoT.

physical entities to the network layer and the application layer through the communication. It includes various IoT terminals composed of IoT chips, modules, sensors, IoT cards, and integrated circuits;

- The network layer mainly has the functions of transmission, routing, and controlling of information, including the extended network, the access network, and the core network. It mainly consists of the Internet, wireless broadband networks, wireless low-speed networks (e.g., ZigBee and Bluetooth), mobile communication networks, and many emerging wireless access techniques such as low power wide area networks (LPWAN) (e.g., long range radio (LoRa) and NB-IoT);

- The communication layer transmits the entire information between layers, which can rely on the Internet and the public telecommunication network, or the industrial private network. It can be considered as the backbone of the IoT architecture;
- The architecture layer, including cloud computing, big data analysis, and other architectures, provides information processing, computing, and other general infrastructure services, capabilities, and resource invocation interfaces for IoT applications. It is the basis for various applications of IoT in various fields;
- The application layer uses the Internet to integrate all devices, sensors, and information through wireless connections, which can realize different IoT applications, such as smart factories, smart homes, smart agriculture, smart transportation, etc [13].

Meanwhile, the IoT network consists of core network, transmission bearer network, and wireless network [14]. The core network needs to support IoT functions and provide IoT service license to meet the writing requirements of users' data. The transmission bearer network uses the existing IP metropolitan area network for data transmission. Among them, the transmission bearing part can be divided into backhaul propagation and fronthaul propagation, where the backhaul propagation refers to the two-layer virtual private network (VPN) docking interaction between the core network and the baseband unit (BBU) through the router, while the forward propagation refers to the direct connection between BBU and radio remote unit (RRU) through optical fiber. The wireless network part is combined with the existing site resources.

Moreover, the network deployment modes can be categorized into independent deployment, protection band deployment, and in-band deployment. The independent deployment mode involves utilizing new bands or unused bands exclusively for deployment purposes. This mode ensures that there is no interference with the existing networks, thereby maintaining optimal network performance. Additionally, the protection band deployment mode primarily utilizes the protection band within the 5G edge spectrum for deployment. By leveraging weak signals, this mode maximizes the utilization of spectrum resources and ensures efficient network operation. Lastly, the in-band deployment mode is predominantly implemented within a specific frequency band of an existing carrier. This mode allows for seamless integration within the carrier's network infrastructure, utilizing the available frequency bands efficiently.

1.3 Characteristics of IoT Signals

In recent years, academia and industry have begun to study the 5G-based IoT. Compared with IoT currently being deployed, 5G-based IoT has more stringent requirements for power consumption, connectivity performance, and coverage. Therefore, it is challenging for traditional technologies to meet these performance requirements. The main challenges are shown as follows:

1) **Signal transmitting with low-time delay:** In order to meet the usage requirements, 5G-based IoT needs to solve the problems of low transmission rate and high delay of 4G networks, which severely limits the applications of IoT and affects the use of various IoT applications. Taking smart home as an example, users need to send information through voice or remote control terminals. After smart home recognizes, receives, and processes the user's information, it will conduct automatic and intelligent operations to achieve the intended purpose of users. All of these processes rely on the high-speed communication network support. Once the network speed is poor or the network time delay is high, the smart home cannot identify, receive, and process the user's information timely, and the entire intelligent operation will have a huge delay, which will seriously affect the actual experience of users. Besides, the automated driving technology, in recent years, also has increasing demands for higher information transmission rate and lower delay of the communication network. If the intelligent vehicle cannot receive terminal instructions or quickly process relevant information of operations due to the low transmission speed, it will cause traffic chaos, and even threaten the life safety of passengers.
2) **Signal detection with low-power users:** IoT continues to evolve from a niche market to a vast network that connects almost every aspect of our lives, thus the power consumption is critical for such a wide range of applications. In the field of IoT, many network devices are sensors, wireless devices, and brakes equipped with data acquisition nodes. In the usual case, these nodes may be power-limited or battery-free devices, which can transmit their data through obtaining and reflecting the energy of the radio frequency (RF) sources. Especially in the industrial installations, these nodes are often placed in areas that are difficult to access or are inaccessible, which means they must operate and transmit data for years on a single button battery.

   Therefore, 5G-based IoT needs to extend the life of the battery in the device by exploiting low-power technologies. It is also necessary to balance the relationship between transmission power, data processing, data storage, and energy consumption, as well as research low-power signal transmission and processing technologies. In addition, new signal detection technologies are supposed to be explored, so that the receiver can successfully detect and receive the signal of terminals.
3) **Signal collision caused by large number of users:** The traditional random access (RA) scheme faces challenges, e.g., up to ten million idle users,

thousands of simultaneous active users, short packet data transmission, and low energy consumption constraints. In the traditional communication systems, RA is a necessary process to establish a wireless link between the terminal and the network. Only after the RA is completed, the data transmission between the terminal and the network can be carried out normally.

However, when the number of users in a cell increases, the performance of RA begins to degrade significantly. Even if the overload control technologies, such as access class barring (ACB), adaptive ACB (A-ACB), extended access barring (EAB) are adopted, with the continuous increase of the number of supported users in the cell, the access success probability of the whole system will be greatly decreased.

4) **Active user sensing of large-scale burst signals:** In a complex IoT ecosystem, there are various subsystems. For example, different drones can be used to serve thousands of industries, e.g., mining and exploration industries can use them for unmanned inspection, media and entertainment industries can use them for aerial photography. Besides, drones can also act as mobile base stations (BS) to provide high-capacity coverage and high-precision positioning services on demand, and large drones play crucial roles in the logistics industry. The common feature of the above 5G-based IoT scenarios is that a large number of terminals carry scattered services, while the device activation is usually sporadic. In each time slot, only a small number of devices are active and can communicate with the BS. Other devices are temporarily dormant to save energy, and they are only activated when triggered by an external event. Therefore, the fundamental challenge of 5G for IoT is that the BS needs to dynamically identify activated devices and complete data transmission in an efficient and timely manner.

1.4 Outline

The remaining themed chapters of this book are introduced as follows: Chapter 2 considers the background of IoT detection, including RA and some signal detection methods. Among them, RA is divided into two types of introductions, namely grant-based and grant-free, which are distinguished by whether to obtain the reserved channel information. Some traditional signal detection methods, such as maximum likelihood (ML), zero forcing (ZF), etc., are also mentioned. Other signal detection algorithms in use, such as variational inference (VI) detection, Markov Chain Monte Carlo (MCMC) detection, etc., are described in detail.

Chapter 3 is concerned with the sparse signal detection for multiple access, with specific technologies like VI detection and compressive sensing (CS) detection.

In this chapter, the algorithm of VI detection is firstly introduced with corresponding analysis of its performance. Then, the theory of CS detection is discussed, including Bayesian CS algorithm and structured subspace pursuit algorithm with the simulation results of them.

In Chapter 4, collided signal detection for multiple access is introduced. With the displayed system model, we analyze the theory of automatic modulation classification-based detection through the simulation results. Accordingly, the simulation results of collided signal detection for multiple access are offered.

Chapter 5 analyses the round-trip delay estimation for collided signals with ML detection, VI detection, and MCMC detection. Firstly, the system model of the round-trip delay estimation for collided signals is preformed. Then, the ML detection is described and we compare it with the VI detection and MCMC detection. Finally, the simulation results of all these detection algorithms are brought together to show their detection performance.

Chapter 6 presents the signal detection for backscatter signal. After giving the explanation of the system model of backscatter signal, we describe the algorithm of the central limit theorem-based detection and analyze its performance by comparing its simulation results with the ML detection.

Chapter 7 firstly explains the concept of non-orthogonal multiple access (NOMA) and adopts the technique of NOMA in IoT systems. Then, this chapter focuses on throughput and power consumption of NOMA systems. Finally, the puzzle of energy efficiency maximization of a full-duplex cooperative NOMA system is investigated.

The purpose of Chapter 8 is to present the signal design for multi-cluster coordination, including multi-cluster coordination with NOMA, constructive interference-aided multi-cluster coordination, and performance analysis. Then, discussions about successive interference cancellation design, device grouping, and power control to achieve multi-cluster coordination with NOMA are included.

In the end, Chapter 9 gives a brief conclusion for this book.

2

Background of IoT Signal Detection

- The aim of this chapter is to introduce random access (RA) in Internet of Things (IoT) systems. Then, we present several well-known detection approaches, including the maximum likelihood (ML), linear, variational inference (VI), Markov chain Monte Carlo (MCMC), and compressive sensing (CS) detectors.

2.1 Random Access

2.1.1 Grant-based Random Access

For grant-based RA, each device needs to collaborate with other devices or base stations (BSs) to obtain a grant for access, which can be further divided into contention-based RA and contention-free RA according to the access resource acquisition methods. For contention-based RA, devices need to initiate access requests on the specified RA resources using the preamble allocated by the BS, where devices obtain the preamble in a competitive way to initiate access requests [15]. Although the contention-free RA allocates specific access resources to access requests that have a high probability of success, the communication between devices should allow the delay caused by the signal collision.

1) **Contention-based random access**: The process of contention-based RA consists of four phases [16], as shown in Fig. 2.1:
   - **Phase 1 – Preamble transmission**: Each device randomly selects an available preamble from the preamble pool upon the arrival of a new RA time slot, and transmits the preamble in the physical random access channel (PRACH) for identification of the device by the BS.
   - **Phase 2 – RA response**: After receiving the preamble transmitted from the device, the BS decodes the received preamble and sends the RA response to the device through the physical downlink shared channel (PDSCH), where the RA response mainly includes the messages contained in the preamble,

IoT Signal Detection, First Edition. Rui Han, Jingjing Wang, Lin Bai, and Jianwei Liu.

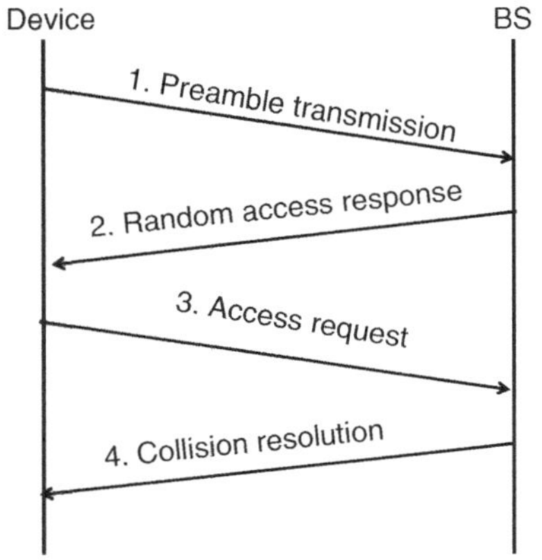

Figure 2.1 The contention-based RA.

the uplink time schedule used for synchronization, the parameters of the retreat, and resource scheduling information, etc.

- **Phase 3 – Connection request**: After receiving the RA response from the BS, the device transmits the connection request on the physical uplink shared channel (PUSCH) allocated to it. If the device does not receive the RA response, it retries the access after a certain period of time.
- **Phase 4 – Collision resolution**: The BS transmits collision resolution information to the device. When the device receives a message corresponding to its identity, the connection between the device and the BS can be established and the device can begin to transmit data. If the device does not receive the corresponding message, it indicates that the device has collides with other devices and needs to resend the access request.

2) **Contention-free random access**: The contention-free RA procedure consists of three phases [16], as shown in Fig. 2.2:

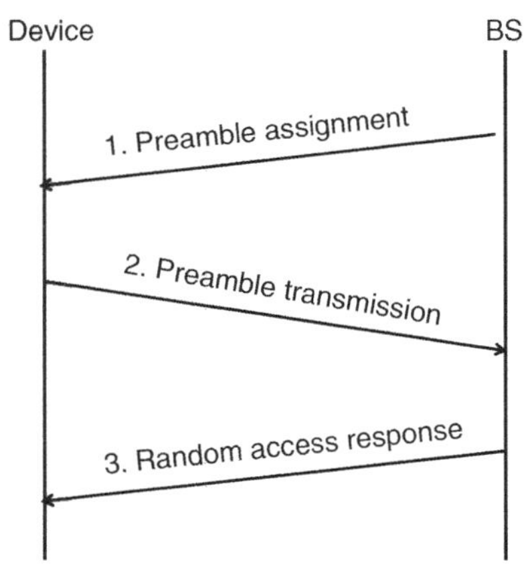

Figure 2.2 The contention-free RA.

- **Phase 1 – Preamble assignment**: The BS assigns the preamble to the device.
- **Phase 2 – Preamble transmission**: The device sends the dedicated preamble on the PRACH.
- **Phase 3 – RA response**: After receiving the preamble from the device, the BS generates the RA response and sends it to the device.

The main advantage of grant-based RA is that it is easy to be operated at the BS and does not require complex access algorithms. However, the grant-based RA has the following drawbacks:

- The number of IoT devices is large while the number of orthogonal preamble is limited, which leads to a high probability that the same preamble is repeatedly selected by multiple devices. Therefore, the access failure probability due to the collision is high.
- Since the grant-based RA scheme requires four handshakes, the signaling overhead is considerable.

To solve the above problems, there have been a number of studies involving the methods of collision resolution in RA protocol, e.g., backoff mechanism, access class barring, time slot access mechanism, priority access mechanism, and other mechanisms, etc [17]. The access class barring mechanism and the backoff mechanism are two common congestion control strategies, which are introduced as follows:

1) **Access class barring**: Access class barring is an important congestion control strategy, which directly controls the traffic by adjusting the probability of RA requests initiated by the grant-based devices on the PRACH. When using the access class barring mechanism, the BS will set an access class barring probability parameter in advance, which is also known as the access class barring factor, and broadcast it before the access time slot. Each device generates a random number and compares it with the access class barring factor. If the generated random number is less than the access class barring factor, the devices will send preamble on the PRACH, otherwise, it will be prevented from access and need to request RA again. Although the access class barring mechanism is promising in congestion control, it increases the time delay and can only be used to alleviate short-term network congestion. Besides, the update speed of the access class barring factor is limited by the broadcast update period, so it cannot timely respond to a large number of access devices in a short time.
2) **Backoff mechanism**: In the backoff mechanism, the BS sets backoff parameters for all devices, including access time and backoff time. When the device fails in the RA process, it first determines whether the maximum number of allowed backoffs has been reached. If not, the waiting time will be selected

according to the backoff indicator for the next access attempt, where the backoff time follows a uniform distribution. Otherwise, the device fails to access and waits for the next activation. However, the collision probability is still high because of the large-scale devices, and the number of access attempts before successful access may be quite high, which increases not only the time delay, but also the energy consumption of the signal transmission. In this case, the traditional grant-based RA is challenging to meet the requirements of the large-scale IoT system.

2.1.2 Grant-free Random Access

Since it is a tough problem for grant-based RA to meet the needs of the large-scale IoT devices, a new RA mechanism should be introduced. Recently, some scholars have proposed grant-free RA to ensure low delay and high reliability of access to IoT devices [18].

As shown in Fig. 2.3, the core idea of grant-free RA is that the activated device directly sends data signals without waiting for the grant of BS after sending the preamble, which can significantly reduce the access delay and the signaling overhead. The crucial assumption in the grant-free transmission is that the BS has completed the ideal active user identity detection, channel estimation, and signal synchronization before data transmission. Thus, how to complete the above detection has become one of the core problems to be solved in the grant-free RA. It can be seen that grant-free RA sacrifices computational complexity to reduce access delay and signaling overhead. Researches on the grant-free RA can be divided into the following categories:

1) **Grant-free random access algorithm based on compressed sensing**: Compressed sensing is a signal processing method further developed by the

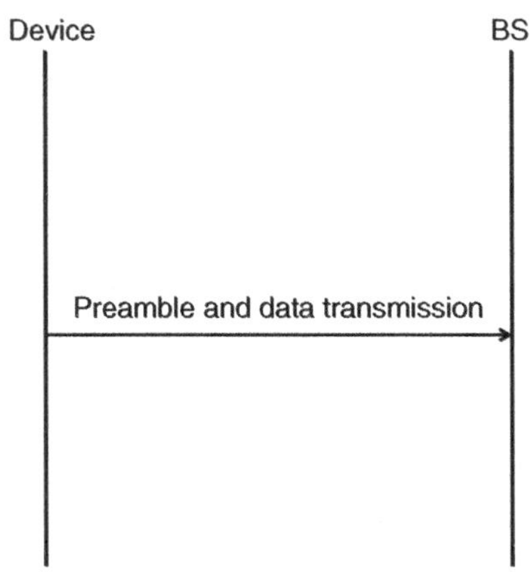

Figure 2.3 The grant-free RA.

Stanford University professor based on the theory of signal sparse decomposition and approximation, which is widely used in digital image processing, communication channel estimation, and other fields. If the signal $\mathbf{x}$ is a sparse signal containing only non-zero elements and the measurement matrix $\mathbf{\Phi}$ is used for observation, the measured signal can be written as

$$\mathbf{y} = \mathbf{\Phi}\mathbf{x} + \mathbf{z}, \tag{2.1}$$

where $\mathbf{z}$ denotes the noise in the transmission process of the observation signal. If the measurement matrix $\mathbf{\Phi}$ satisfies the criteria of restricted isometry property (RIP), the receiver can accurately reconstruct the sparse signal $\mathbf{x}$ based on the signal $\mathbf{y}$, which is shown in Fig. 2.4.

In general, the complexity of the massive grant-free RA is proportional to the total number of devices. Therefore, although the above-mentioned grant-free RA based on compressed sensing can reduce the signaling overhead in IoT, the high computational complexity of these algorithms is still an urgent problem to be solved.

2) **Grant-free random access algorithm based on statistics**: In general, the number of allowed activated devices is proportional to the quadratic of the preamble length, which severely limits the number of activated devices that can be held in the system. If the BS is equipped with large-scale antennas, it is of great research value to design grant-free RA based on the covariance matrix of received signals. However, large antenna arrays lead to high array dimension and hardware cost, and centralized processing of massive array signals may become impractical. Thus, distributed multiple-input multiple-output (MIMO) maybe a feasible method to support large-scale RA in the future [19].

Although the above access algorithms based on compressed sensing and statistical information provide some specific solutions for grant-free RA, the

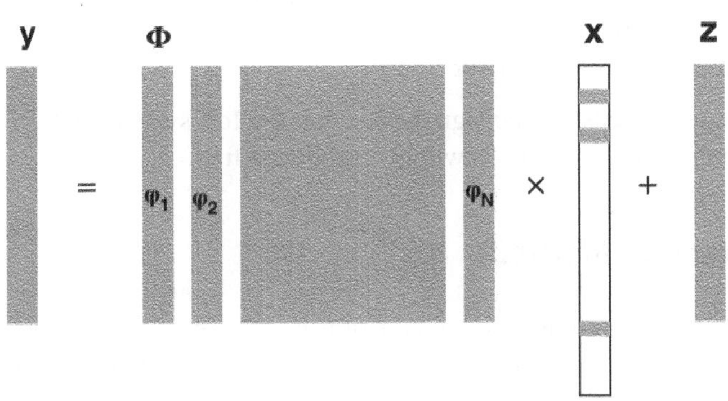

Figure 2.4 The compressed sensing.

characteristics of IoT devices (e.g., high-dense and diversification) present some new challenges, which are summarized as follows:

- **Signal asynchronous**: The models in most of the researches assume that devices can achieve ideal signal synchronization. However, in the actual grant-free RA, different devices have different timing deviations, which causes the preambles received by the BS from the devices to be not fully synchronized. The phenomenon of preamble asynchronism can cause the performance of active user detection algorithm to deteriorate sharply or even fail completely, which is a major challenge in the practical application of the grant-free RA. The research of large-scale RA in asynchronous scenarios is still in the exploratory stage, and there are few related studies.
- **Preamble shortage**: The RA scheme needs to assign the corresponding preamble to potential devices so that they can be activated and connected at any time. However, since the number of active devices is large, assigning a dedicated preamble to each device causes the sequence length of preamble to be excessively long, which reduces the transmission efficiency. In addition, due to the limitation of the coherence time of channels, the preamble cannot be designed too long. Therefore, how to design an RA algorithm to reduce the preamble overhead is an urgent problem to be solved.
- **High computational complexity**: The larger the number of antennas, the higher the detection accuracy of grant-free RA. However. the computational complexity of the algorithm will increase sharply. The technique of multi-antenna is an important technology to support future large-scale IoT applications. Due to the deployment of large-scale antenna arrays and the access of large-scale IoT devices, the signal received by the BS is ultra-high dimensional in the spatial domain and frequency domain, which makes the operation complexity extremely high and affects the practicability of the algorithm. Moreover, the high computational complexity leads to the long processing time, which results in the high access delay. Therefore, how to design a low complexity algorithm for jointly activated device detection and channel estimation is a troublesome issue.

In this book, we mainly focus on the signal detection for IoT systems, where some traditional signal detection methods will be introduced first.

2.2 Signal Detection Methods

In this section, we first introduce the system model of IoT communications. Then, we talk about the fundamental signal detection methods, which are ML detection, zero forcing (ZF) detection, minimum mean square error (MMSE) detection, and the modern signal detection methods, such as MCMC detection, VI detection, and

CS detection. These methods focus on the detection of a group of signals, i.e., the signal vector, instead of a single signal.

2.2.1 System Model

As the proliferation of IoT devices escalated in forthcoming wireless networks, the next generation multiple access (NGMA) schemes are being contemplated to facilitate extensive connectivity for a vast multitude of devices. In IoT scenarios, it is expected to support a tremendous amount of communicating machines, thus machine-type communication (MTC) becomes essential.

In order to establish communication with a BS, an IoT device must initiate a procedure of RA. This involves transmitting a preamble, randomly chosen from a predefined set, through a shared PRACH. In scenarios where a large number of devices are attempting to access the network simultaneously, it is common for multiple IoT devices to select the same preamble, resulting in preamble collisions. Unfortunately, when collisions occur, the data packet transmitted by the devices involved may fail to be decoded successfully at the BS. Therefore, mitigating congestion and overload on the PRACH is of utmost significance, particularly in the context of supporting massive connectivity to facilitate MTC.

Recent years have witnessed a growing academic interest in MIMO technique due to its crucial importance in providing extensive spatial freedom and improving spectral efficiency, MIMO has become a significant technique. As a result, many IoT devices are now equipped with multiple antennas, and a variety of transmission schemes have been proposed to exploit the capabilities of MIMO channels. These schemes aim to achieve a harmonious balance between the advantages of diversity and multiplexing.

A typical MIMO system is shown in Fig. 2.5, and its transmission model can be written as

$$\mathbf{r} = \mathbf{H}\mathbf{s} + \mathbf{n}, \tag{2.2}$$

where $\mathbf{r}$ is a vector of N dimensions, which represents the received signal; $\mathbf{s}$ is a vector of M dimensions, which represents the signal to be detected; $\mathbf{H}$ is a $N \times M$ channel matrix, and $\mathbf{n} \sim \mathcal{CN}(\mathbf{0}, N_0\mathbf{I})$ denotes the noise vector with N dimensions.

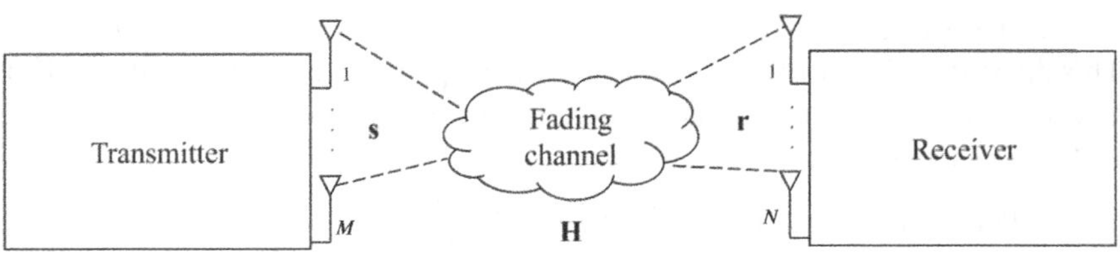

Figure 2.5 A $M \times N$ MIMO system.

Let $\mathbf{s} = [s_1 \ldots s_m]^{\mathrm{T}}$, where $s_m \in S$ is a discrete random variable. The set S represents the signal constellation. We denote $\Pr(s_m)$ as the a priori probability of s_m, and assume independence among all s_m. To estimate $\mathbf{s}$, we can address the maximum a posteriori probability (MAP) problem illustrated as

$$\hat{\mathbf{b}}_{\text{map}} = \underset{\mathbf{s} \in S^{\mathcal{M}}}{\operatorname{argmax}} \Pr(\mathbf{s}|\mathbf{r}), \tag{2.3}$$

where $\Pr(\mathbf{s} \mid \mathbf{r})$ represents the a posteriori probability of $\mathbf{s}$.

The problem outlined in Eq. (2.3) is known as a MIMO detection problem. Generally, finding the MAP solution for this problem is challenging due to its computational complexity, which may be unfeasible. Consequently, various low-complexity approaches have been explored to tackle the following problems.

2.2.2 ML Detection

ML detection searches all candidate vectors exhaustively and selects the largest possible vector with the smallest error probability as the received signal vector [20]. Generally, the method calculates the Euclidean distance of the received signal vector, and finds out the smallest Euclidean distance among all possibilities and the corresponding possible vector is chosen as the target transmission signal vector. In this case, the ML detection has the best detection performance.

Let $f(\mathbf{r}|\mathbf{s})$ denote the likelihood function of $\mathbf{s}$ for a given received signal vector $\mathbf{r}$, then the estimated vector of ML detection can be expressed as

$$\begin{aligned} \hat{\mathbf{s}}_{\text{ML}} &= \underset{\mathbf{s} \in \Omega}{\operatorname{argmax}} \ f(\mathbf{r}|\mathbf{s}) \\ &= \underset{\mathbf{s} \in \Omega}{\operatorname{argmin}} \ \|\mathbf{r} - \mathbf{H}\mathbf{s}\|_2^2, \end{aligned} \tag{2.4}$$

where Ω is the set of all signal candidates on the constellation and $\|\cdot\|_2$ denotes the 2-norm.

All transmit signal vectors in the constellation will be traversed to deal with Eq. (2.4). If the transmitter has M antennas and the modulation order is K, then the traversal search space is $|\Omega| = 2^{MK}$. In other words, the computational complexity increases exponentially with the number of antennas and the modulation order. Therefore, although ML detection has good performance, it is generally not applied in massive MIMO detection. Through the ML detection, the fully utilized spatial diversity gain is achieved, which equals the number of receive antennas N. This diversity gain is derived by considering the pairwise error probability (PEP). If $\mathbf{s}_1$ denotes the transmit signal, $\mathbf{s}_2$ denotes the erroneously detected signal, the PEP can be expressed as

$$\begin{aligned} P(\mathbf{s}_1 \to \mathbf{s}_2) &= \Pr\left(\|\mathbf{r} - \mathbf{H}\mathbf{s}_2\|_2^2 \leq \|\mathbf{r} - \mathbf{H}\mathbf{s}_1\|_2^2 \right) \\ &= \Pr\left(\|\mathbf{H}(\mathbf{s}_1 - \mathbf{s}_2) + \mathbf{n}\|_2^2 \leq \|\mathbf{n}\|_2^2 \right) \\ &= \Pr\left(\|\mathbf{H}\mathbf{d}\|_2^2 \leq -2\Re\left(\mathbf{n}^{\mathrm{H}}\mathbf{H}\mathbf{d}\right) \right), \end{aligned} \tag{2.5}$$

where $\mathbf{d} = \mathbf{s}_1 - \mathbf{s}_2$ and $\Re(\cdot)$ denotes the real-part operation. The noise vector $\mathbf{n}$ is a circularly symmetric complex Gaussian (CSCG) random vector, then $\mathbf{n}^{\mathrm{H}}\mathbf{Hd}$ also becomes a CSCG random vector, which has the following properties:

$$\begin{aligned}
&\mathbb{E}\left[\mathbf{n}^{\mathrm{H}}\mathbf{Hd}\right] = \mathbf{0},\\
&\mathbb{E}\left[\mathbf{n}^{\mathrm{H}}\mathbf{Hdd}^{\mathrm{H}}\mathbf{H}^{\mathrm{H}}\mathbf{n}\right] = \mathbb{E}\left[\mathrm{Tr}\left(\mathbf{H}^{\mathrm{H}}\mathbf{nn}^{\mathrm{H}}\mathbf{Hdd}^{\mathrm{H}}\right)\right] = N_0\mathrm{Tr}\left(\mathbf{H}^{\mathrm{H}}\mathbf{Hdd}^{\mathrm{H}}\right),\\
&\mathbb{E}\left[\mathbf{n}^{\mathrm{H}}\mathbf{Hdd}^{\mathrm{H}}\mathbf{H}^{\mathrm{H}}\mathbf{n}^*\right] = \mathbb{E}\left[\mathrm{Tr}\left(\mathbf{H}^{\mathrm{H}}\mathbf{n}^*\mathbf{n}^{\mathrm{H}}\mathbf{Hdd}^{\mathrm{T}}\right)\right] = \mathbf{0},
\end{aligned} \tag{2.6}$$

where $\mathbb{E}[\cdot]$ represents the expectation for the element in brackets and $\mathrm{Tr}(\cdot)$ denotes the trace of the matrix in brackets.

According to Eq. (2.6), it can be obtained that

$$\mathbf{n}^{\mathrm{H}}\mathbf{Hd} \sim \mathcal{N}\left(0, \frac{N_0}{2}\mathrm{Tr}(\mathbf{H}^{\mathrm{H}}\mathbf{Hdd}^{\mathrm{H}})\right). \tag{2.7}$$

Since for any matrix $\mathbf{A}$,

$$\mathrm{Tr}\left(\mathbf{AA}^{\mathrm{H}}\right) = \mathrm{Tr}\left(\mathbf{A}^{\mathrm{H}}\mathbf{A}\right) = \|\mathbf{A}\|_{\mathrm{F}}^2, \tag{2.8}$$

the variance of $\mathbf{n}^{\mathrm{H}}\mathbf{Hd}$ becomes

$$\begin{aligned}
\frac{N_0}{2}\mathrm{Tr}(\mathbf{H}^{\mathrm{H}}\mathbf{Hdd}^{\mathrm{H}}) &= \frac{N_0}{2}\|\mathbf{Hd}\|_{\mathrm{F}}^2\\
&= \frac{N_0}{2}\|\mathbf{Hd}\|_2^2,
\end{aligned} \tag{2.9}$$

where $\|\cdot\|_{\mathrm{F}}$ denotes the Frobenius norm. Then, the PEP can be expressed by the Q-function [21] as

$$\begin{aligned}
P(\mathbf{s}_1 \to \mathbf{s}_2) &= \Pr\left(\|\mathbf{Hd}\|_2^2 \le -2\mathbf{n}^{\mathrm{H}}\mathbf{Hd}\right)\\
&= Q\left(\sqrt{\frac{\|\mathbf{Hd}\|_2^2}{2N_0}}\right),
\end{aligned} \tag{2.10}$$

where $Q(\cdot) = \int_x^{+\infty} \frac{1}{\sqrt{2\pi}} \exp\left(-\frac{1}{2}t^2\right) dt$. According to Eq. (2.10), the upper bound of PEP can be computed as

$$\begin{aligned}
P(\mathbf{s}_1 \to \mathbf{s}_2) &\le Q\left(\sqrt{\frac{\|\mathbf{Hd}_{\min}\|_{\mathrm{F}}^2}{2N_0}}\right)\\
&\le \sum_{\mathbf{d}\in D, \mathbf{d}\neq\mathbf{0}} Q\left(\sqrt{\frac{\|\mathbf{Hd}\|_2^2}{2N_0}}\right),
\end{aligned} \tag{2.11}$$

where $D = \left\{\mathbf{d} = \mathbf{s}_1 - \mathbf{s}_2 | \mathbf{s}_1, \mathbf{s}_2 \in \Omega\right\}$, and

$$\mathbf{d}_{\min} = \underset{\mathbf{d}\in D, \mathbf{d}\neq\mathbf{0}}{\mathrm{argmin}} \|\mathbf{Hd}\|_2^2. \tag{2.12}$$

According to the Chernoff boundary, the average PEP's upper bound can be given by

$$\begin{aligned}\mathbb{E}\left[P(\mathbf{s}_1 \rightarrow \mathbf{s}_2)\right] &\leq \sum_{\mathbf{d}\in\mathcal{D},\mathbf{d}\neq\mathbf{0}} \mathbb{E}\left[\exp\left(-\frac{\|\mathbf{Hd}\|_2^2}{2N_0}\right)\right] \\ &\leq \sum_{\mathbf{d}\in\mathcal{D},\mathbf{d}\neq\mathbf{0}} \det\left(\mathbf{I}+\frac{\sigma_h^2}{4N_0}\mathbf{dd}^{\mathrm{H}}\right)^{-N},\end{aligned} \tag{2.13}$$

where $\det(\cdot)$ denotes the determinant of the matrix, and σ_h^2 represents the variance of the channel matrix $\mathbf{H}$. Equation (2.13) shows that a complete receive diversity gain of N can be achieved by the ML detection with the given system model.

The process of ML detection is presented in Algorithm 2.1.

Algorithm 2.1 ML Detection

Input: $\mathbf{r}, \mathbf{H}, \mathbf{M}, \mathbf{K}$
Output: $\hat{\mathbf{s}}_{\mathrm{ML}}$

```
Initialize: Ω
for i∈{1,···,2^{MK}} do
    for s_i ∈ Ω do
        Compute || r − Hs_i ||_2^2;
        if || r − Hs_i ||_2^2 < || r − Hs_{i−1} ||_2^2 then
            Update ŝ_ML according to Eq. (2.4);
        end if
    end for
end for
Return ŝ_ML;
```

Example 2.1 Consider a 2×2 MIMO system with 64 quadrature amplitude modulation (QAM) modulation method. Assume that the channel matrix is $\mathbf{H}=\begin{bmatrix}0.63+1.53i & -2.12+1.08i\\ 0.47-0.89i & -0.39+0.52i\end{bmatrix}$, and the noise matrix is $\mathbf{n}=\begin{bmatrix}0.03+0.04i\\ -0.04-0.02i\end{bmatrix}$. With the signal vector $\mathbf{s}=\begin{bmatrix}3+5i\\ -1-7i\end{bmatrix}$ transmitted in 64-QAM, the received signal is

$$\begin{aligned}\mathbf{r} &= \mathbf{Hs}+\mathbf{n} \\ &= \begin{bmatrix}3.95+21.54i\\ 9.85+1.87i\end{bmatrix}.\end{aligned} \tag{2.14}$$

Let

$$d_{\mathrm{ML}} = \|\mathbf{r}-\mathbf{Hs}_i\|_2^2, \tag{2.15}$$

where $\mathbf{s}_l$ is the candidate vector. Note that the power is normalized. We know that there are 4096 candidate vectors in 64-QAM, then we substitute the candidate vector into Eq. (2.15) and find the minimum value of d_{ML}. Through substituting every candidate vector and calculating, the minimum value of d_{ML} is 0.0045 and the corresponding vector is $\mathbf{s}_{\mathrm{ML}} = \begin{bmatrix} 3+5i \\ -1-7i \end{bmatrix}$.

For a more generalized scenario, Fig. 2.6 demonstrates the constellation diagram of the estimated signal obtained by ML detection in a 64-QAM communication system with the same response matrix. Therefore, this shows that the ML detector can correctly detect the signal, but because of its exhaustive search, it trades high time complexity for good detection performance.

In order to numerically analyze the performance of the ML detector, we will simulate this detection method, where the bit error rate (BER) is considered as a performance metric in simulation at a certain signal-to-noise ratio (SNR).

In Fig. 2.7, we show the BER of ML detector in 4-QAM 2×2 MIMO system. What stands out in this figure is that the BER of ML detector drops from 0.3085 to 4×10^{-7}, when SNR increases from 0 to 20 dB.

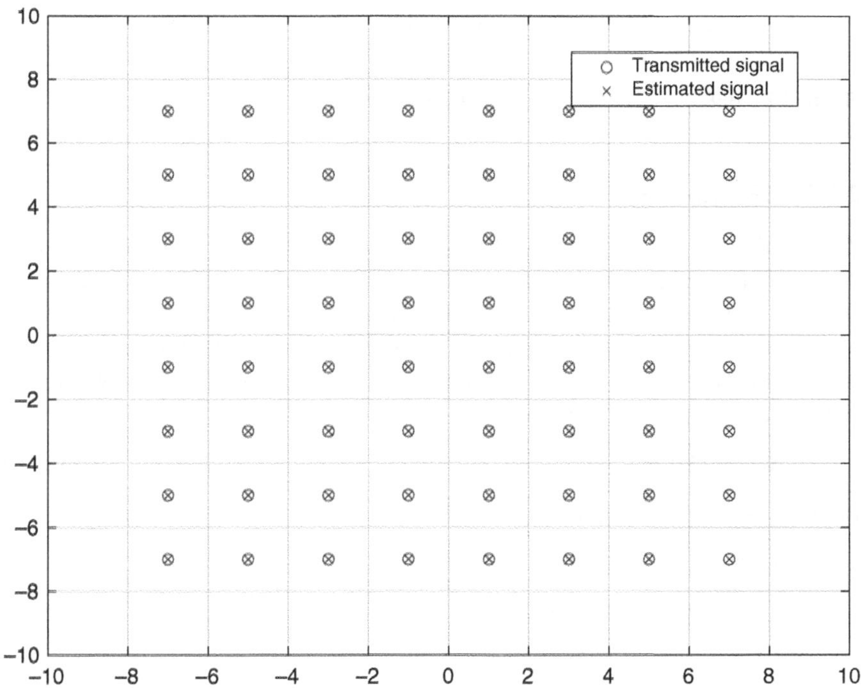

Figure 2.6 The constellation of 64-QAM system with ML.

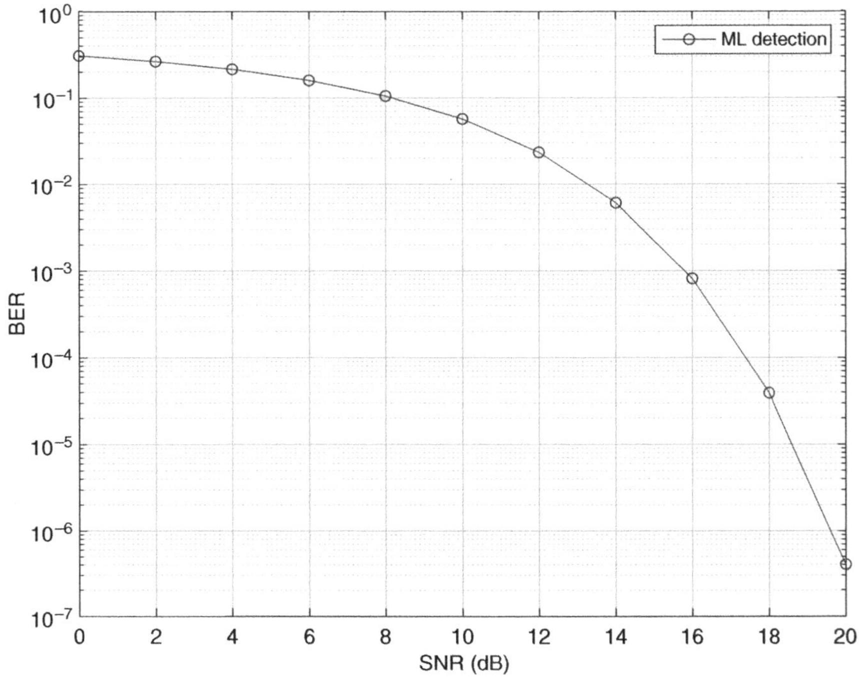

Figure 2.7 BER of ML detector in 4-QAM 2 × 2 MIMO system.

2.2.3 ZF Detection

ZF detection is one of the linear detection methods. The signal received by a linear detector is filtered by a well-designed matrix which enables the transmitted multiple signals to be detected separately without involving interference. The main idea of ZF detection can be summarized as using linear transformation to reduce the mutual interference between diverse antennas, to obtain useful signals, where the estimator matrix can be expressed as

$$\mathbf{W}_{\mathrm{ZF}} = \left(\mathbf{H}^{\mathrm{H}}\mathbf{H}\right)^{-1}\mathbf{H}^{\mathrm{H}}. \tag{2.16}$$

Besides, the estimated signal can be given by

$$\begin{aligned}\hat{\mathbf{s}}_{\mathrm{ZF}} &= \mathbf{W}_{\mathrm{ZF}}\mathbf{r} \\ &= \left(\mathbf{H}^{\mathrm{H}}\mathbf{H}\right)^{-1}\mathbf{H}^{\mathrm{H}}\left(\mathbf{H}\mathbf{s}+\mathbf{n}\right) \\ &= \mathbf{s} + \left(\mathbf{H}^{\mathrm{H}}\mathbf{H}\right)^{-1}\mathbf{H}^{\mathrm{H}}\mathbf{n}.\end{aligned} \tag{2.17}$$

Thus, the covariance matrix of the estimated error of the transmit signal is

$$\begin{aligned}\boldsymbol{\varphi}_{\mathrm{ZF}} &= \mathbb{E}\left\{\left(\hat{\mathbf{s}}_{\mathrm{ZF}}-\mathbf{s}\right)\left(\hat{\mathbf{s}}_{\mathrm{ZF}}-\mathbf{s}\right)^{\mathrm{T}}\right\}\\ &= \sigma_n^2\mathbf{W}_{\mathrm{ZF}}\left(\mathbf{W}_{\mathrm{ZF}}\right)^{\mathrm{T}}\\ &= \sigma_n^2\left(\mathbf{H}^{\mathrm{T}}\mathbf{H}\right)^{-1},\end{aligned} \tag{2.18}$$

where σ_n^2 denotes the variance of noise.

It can be seen that the operation of ZF detection is simple, but it ignores the influence of noise. In particular, the noise is amplified after being filtered and detected by ZF estimator, and the estimation error of the transmit signal becomes larger. Thus, ZF detection cannot offer a satisfied BER performance when SNR is low.

The process of ZF detection is presented in Algorithm 2.2.

Algorithm 2.2 ZF Detection

Input: $\mathbf{r}$, $\mathbf{H}$, $\mathbf{s}$

Output: The reconstructed signal

1: Compute the estimator matrix, $\mathbf{W}_{\mathrm{ZF}}$;
2: Compute the estimated signal, $\hat{\mathbf{s}}_{\mathrm{ZF}}$;
3: Compute the covariance matrix of the estimated error of the transmit signal, $\boldsymbol{\varphi}_{\mathrm{ZF}}$;
4: Decide the reconstructed signal according to decision threshold and $\hat{\mathbf{s}}_{\mathrm{ZF}}$;

Example 2.2 Consider a 2×2 MIMO system with 64-QAM modulation method. Assume that the channel matrix is $\mathbf{H} = \begin{bmatrix} 0.82+1.53i & 0.59+1.13i \\ 0.63-0.21i & 0.19-0.69i \end{bmatrix}$, and the noise matrix is $\mathbf{n} = \begin{bmatrix} 0.04+0.01i \\ -0.04-0.03i \end{bmatrix}$. With the signal vector $\mathbf{s} = \begin{bmatrix} -1-3i \\ 3+i \end{bmatrix}$ transmitted in 64-QAM, the received signal is

$$\begin{aligned}\mathbf{r} &= \mathbf{Hs}+\mathbf{n}\\ &= \begin{bmatrix} 4.45 \\ -0.04-3.59i \end{bmatrix}.\end{aligned} \tag{2.19}$$

The estimator matrix is

$$\begin{aligned}\mathbf{W}_{\mathrm{ZF}} &= \left(\mathbf{H}^{\mathrm{H}}\mathbf{H}\right)^{-1}\mathbf{H}^{\mathrm{H}}\\ &= \begin{bmatrix} 0.64-0.23i & 0.60-1.07i \\ -0.51-0.38i & -0.75+1.47i \end{bmatrix}.\end{aligned} \tag{2.20}$$

The estimated signal becomes

$$\begin{aligned}\hat{\mathbf{s}}_{\mathrm{ZF}} &= \mathbf{W}_{\mathrm{ZF}}\mathbf{r} \\ &= \begin{bmatrix} -1.03 - 2.98i \\ 3.06 + 0.94i \end{bmatrix}.\end{aligned} \tag{2.21}$$

Then it can be determined that the estimated signal vector is $\begin{bmatrix} -1 - 3i \\ 3 + i \end{bmatrix}$ by hard judgment. Therefore, this shows that ZF detection can correctly detect signal.

Besides, the covariance matrix of the estimated error of the transmit signal is

$$\begin{aligned}\varphi_{\mathrm{ZF}} &= \sigma_n^2\left(\mathbf{H}^{\mathrm{T}}\mathbf{H}\right)^{-1} \\ &= \begin{bmatrix} -0.0077i & -0.0089 + 0.0013i \\ -0.0089 - 0.0013i & 0.0125 \end{bmatrix},\end{aligned} \tag{2.22}$$

from which we can see that the estimated error of the transmit signal is small.

To give a more generalized scenario, Fig. 2.8 reveals the constellation diagram of the estimated signal obtained by ZF detection in a 64-QAM communication system with the same response matrix. It is apparent from this figure that ZF detection can correctly detect signal.

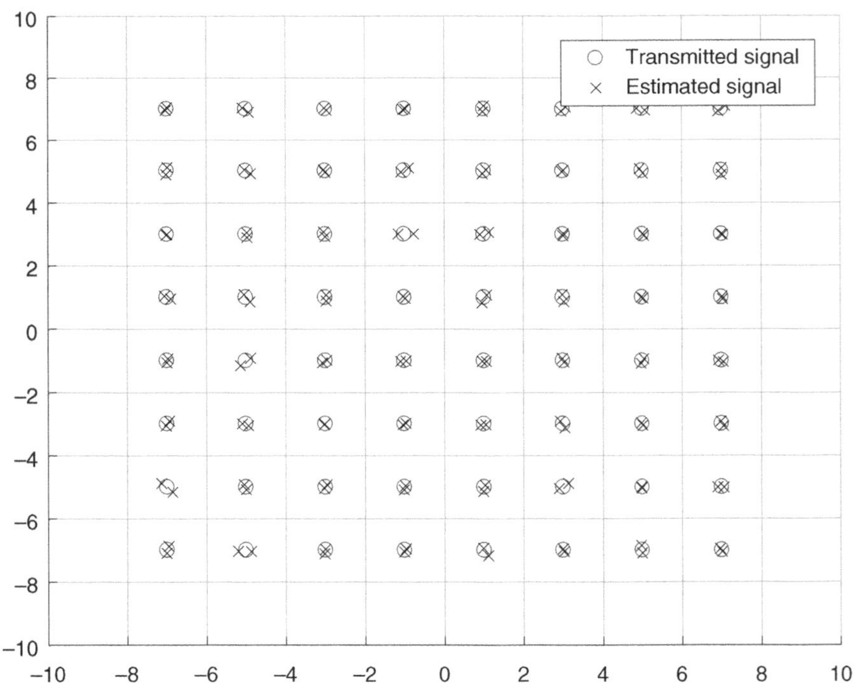

Figure 2.8 The constellation of 64-QAM system with ZF.

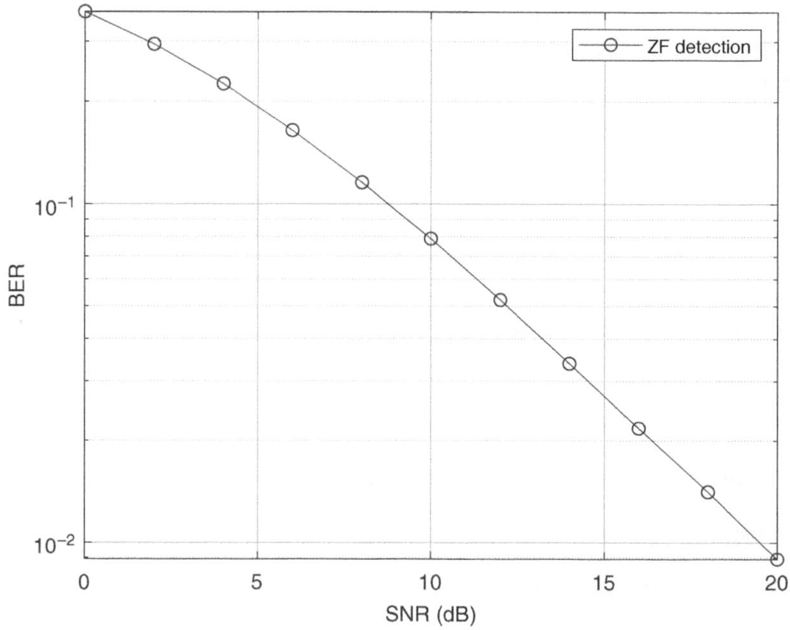

Figure 2.9 BER of ZF detector in 4-QAM 2×2 MIMO system.

In order to numerically analyze the performance of ZF detector, we will simulate this detection method. Figure 2.9 shows the BER of ZF detector in 4-QAM 2×2 MIMO system. There is a clear trend in Fig. 2.9 that the BER of ZF detector drops from 0.3671 to 8.922×10^{-3}, when SNR increasing from 0 to 20 dB.

2.2.4 MMSE Detection

The MMSE detection makes up for the defect of noise amplification in the ZF detection. The main idea of MMSE detection is to calculate the mean square error between the detected signal $\hat{\mathbf{s}}$ and the actual transmitted signal $\mathbf{s}$, to identify the estimated value that minimizes the mean square error as the detection result.

There are mainly two aspects during the MMSE detection process: channel estimation and signal estimation, as shown in Fig. 2.10.

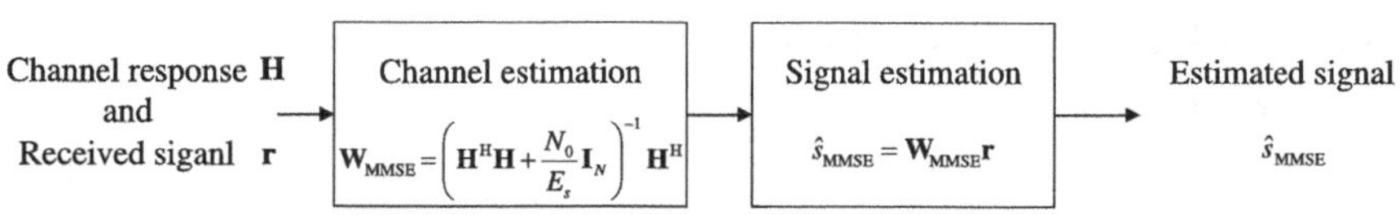

Figure 2.10 The process of MMSE detection.

In the first step, the scheme to obtain the MMSE filter is to minimize the mean-square error (MSE). The solution of this minimization problem, as well as the corresponding MMSE estimator matrix $\mathbf{W}_{\text{MMSE}}$, can be computed as

$$\begin{aligned}\mathbf{W}_{\text{MMSE}} &= \underset{\mathbf{W}}{\operatorname{argmin}}\ \mathbb{E}\left(\|\mathbf{Wr}-\mathbf{s}\|^2\right)\\ &= (\mathbb{E}[\mathbf{rr}^{\text{H}}])^{-1}\mathbb{E}[\mathbf{rs}^{\text{H}}]\\ &= \left(\mathbf{H}^{\text{H}}\mathbf{H}+\frac{N_0}{E_s}\mathbf{I}_N\right)^{-1}\mathbf{H}^{\text{H}},\end{aligned} \tag{2.23}$$

where $\mathbf{W}$ denotes the MMSE estimator matrix, $\mathbf{I}_N$ is the N-order identity matrix, and E_s represents the symbol energy.

In the second step, the resulting estimated signal vector is

$$\hat{\mathbf{s}}_{\text{MMSE}} = \mathbf{W}_{\text{MMSE}}\mathbf{r} = \left(\mathbf{H}^{\text{H}}\mathbf{H}+\frac{N_0}{E_s}\mathbf{I}_N\right)^{-1}\mathbf{H}^{\text{H}}\mathbf{r}. \tag{2.24}$$

The covariance matrix of the estimated error of the transmit signal can be given by

$$\begin{aligned}\varphi_{\text{MMSE}} &= \mathbb{E}\left(\left(\hat{\mathbf{s}}_{\text{MMSE}}-\mathbf{s}\right)\left(\hat{\mathbf{s}}_{\text{MMSE}}-\mathbf{s}\right)^{\text{H}}\right)\\ &= \left(\mathbf{H}^{\text{H}}\mathbf{H}+\frac{N_0}{E_s}\mathbf{I}_N\right)^{-1}.\end{aligned} \tag{2.25}$$

Compared with Eq. (2.18), the MMSE detection suppresses noise amplification and the estimation error of the transmit signal is smaller.

The specific steps of MMSE detection are shown in Algorithm 2.3.

Algorithm 2.3 MMSE Detection

Input: $\mathbf{r}$, $\mathbf{H}$, N_0

Output: estimated signal $\hat{\mathbf{s}}_{\text{MMSE}}$

1: Compute the estimator matrix, $\mathbf{W}_{\text{MMSE}} = (\mathbf{H}^{\text{H}} + \frac{N_0}{E_s}\mathbf{I}_N)^{-1}\mathbf{H}^{\text{H}}$;
2: Compute the estimated signal, $\hat{\mathbf{s}}_{\text{MMSE}} = \mathbf{W}_{\text{MMSE}}\mathbf{r}$;
3: Decide the reconstructed signal according to decision threshold and $\hat{\mathbf{s}}_{\text{MMSE}}$;

Example 2.3 Consider a 2×2 MIMO system with 64-QAM modulation method. Assume that the channel matrix is $\mathbf{H} = \begin{bmatrix} 0.89-1.15i & -1.07-0.81i \\ -2.94+1.44i & 0.33-0.75i \end{bmatrix}$, and the noise matrix is $\mathbf{n} = \begin{bmatrix} 0.03+0.03i \\ -0.04+0.02i \end{bmatrix}$. With the signal vector $\mathbf{s} = \begin{bmatrix} -3+3i \\ 1+7i \end{bmatrix}$ transmitted in 64-QAM, the received signal is

$$\begin{aligned}\mathbf{r} &= \mathbf{Hs}+\mathbf{n}\\ &= \begin{bmatrix} 5.41-2.15i \\ 10.04+11.56i \end{bmatrix}.\end{aligned} \tag{2.26}$$

The estimator matrix is

$$\mathbf{W}_{\text{MMSE}} = \left(\mathbf{H}^{\text{H}}\mathbf{H} + \frac{N_0}{E_s}\mathbf{I}_N\right)^{-1}\mathbf{H}^{\text{H}} = \begin{bmatrix} -0.01 + 0.16i & -0.25 - 0.07i \\ -0.42 + 0.46i & -0.08 + 0.27i \end{bmatrix}. \tag{2.27}$$

The estimated signal is

$$\hat{\mathbf{s}}_{\text{MMSE}} = \mathbf{W}_{\text{MMSE}}\mathbf{r} = \begin{bmatrix} -2.99 + 3.00i \\ 0.97 + 6.99i \end{bmatrix}. \tag{2.28}$$

Similarly, the vector $\begin{bmatrix} -3 + 3i \\ 1 + 7i \end{bmatrix}$ is decided as the estimated signal vector by hard judgment.

For a more generalized scenario, Fig. 2.11 presents the constellation diagram of the estimated signal obtained by MMSE detection in a 64-QAM communication system with the same response matrix. Therefore, it can be proved that MMSE detection can correctly detect signal.

To conduct a numerical analysis of the MMSE detector's performance, the simulation of this detection method is provided. In Fig. 2.12, we show the BER

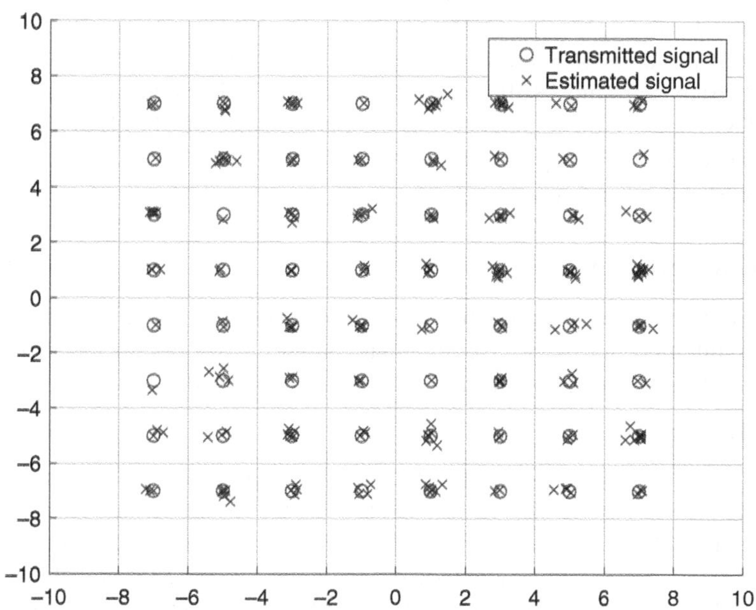

Figure 2.11 The constellation of 64-QAM system with MMSE.

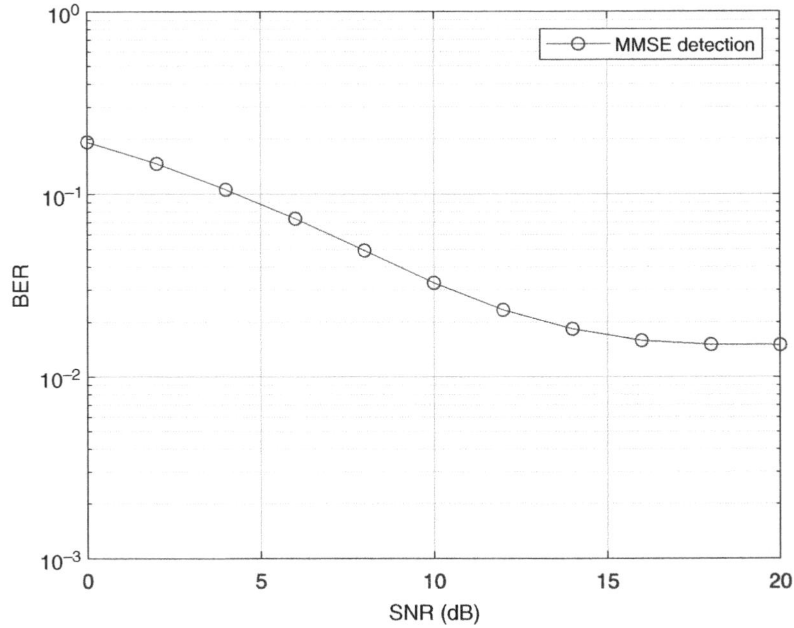

Figure 2.12 BER of MMSE detector in 4-QAM system.

of MMSE detector in 4-QAM 2×2 MIMO system with SNR from 0 to 20 dB. This figure illustrates that as SNR increases, the BER of MMSE detector decreases. When SNR reaches around 18 dB, the BER is approximately constant with the increased SNR. This result highlights that the performance of MMSE detector can be enhanced by increasing SNR. However, the performance is approximately stable when SNR reaches a specific value.

2.2.5 MCMC Detection

To state the system model, suppose that a $M \times N$ MIMO system is equipped with M transmit antennas and N receive antennas. Based on the system model in Section 2.2.1, the signal received by the receiver can be assumed as

$$\mathbf{r} = \mathbf{H}\mathbf{s} + \mathbf{n}, \tag{2.29}$$

where $\mathbf{H} \in \mathbb{C}^{M\times N}$, $\mathbf{s} = (s_1 \cdots s_M)^{\mathrm{T}}$ denote the $M \times N$ channel matrix and the transmitted signal vector, respectively, and $\mathbf{n} \sim \mathcal{CN}(\mathbf{0}, N_0\mathbf{I})$ represents the noise vector.

The detection algorithm based on MCMC obtains the list of transmitted sequences by statistical sampling method, and estimates the posterior probability of each encoded bit based on statistical principle [22]. Since the performance of MCMC detection is determined by the number of samples and independent of the dimension of the random variable to be estimated, it can effectively solve the problem that the algorithm complexity increases exponentially with the number of transmit antennas and modulation orders. The MCMC algorithm was first applied to physics by Metropolis in 1950 [23]. Its main idea is to obtain a sample of the desired distribution $\pi(\mathbf{s})$ by means of the Markov chain. By choosing the appropriate transfer kernel, the sampling value of the Markov chain with any initial distribution can approach to the expected distribution $\pi(\mathbf{s})$ after enough sampling times.

In MIMO systems, each transmitted symbol sequence $\mathbf{s}$ is defined as a state of the Markov chain. Since the number of states of the Markov chain increases exponentially with the dimension of $\mathbf{s}$ in MIMO systems, Gibbs Sampler is applied to reduce the sampling complexity, which is one of the most practical MCMC sampling methods [24]. Let $p_e(s)$ denote the prior probability of s, $s^{(n)}$ represents the variable at the nth iteration. The transfer kernel is obtained as

$$p(s_i|s_{i-}^{(n)}, r) = \frac{p(r|s_i, s_{i-}^{(n)})p_e(s_i)}{\sum\limits_{s_i} p(r|s_i, s_{i-}^{(n)})p_e(s_i)}, \tag{2.30}$$

where $s_{i-}^{(n)} = (s_1^{(n)}, \cdots s_{i-1}^{(n)}, s_{i+1}^{(n)} \cdots, s_M^{(n)})$ denotes the complement set of $s_i^{(n)}$. Analyzing Eq. (2.30), we can obtain that

$$p(s_i|r, s_{i-}^{(n)}) = \frac{p(r|s_i, s_{i-}^{(n)})p_e(s_i)}{p(r|s_{i-}^{(n)})}, \tag{2.31}$$

$$p(s_{i-}^{(n)}|r) = \frac{p(r|s_{i-}^{(n)})p_e(s_{i-}^{(n)})}{p(r)}. \tag{2.32}$$

From $\sum_{s_i} p(s_i|r) = 1$, it can be derived that

$$p\left(s_i = \omega_m \mid r\right) = \frac{\sum_{n=1}^{N} p\left(r \mid s_i = \omega_m, s_{i-}^{(n)}\right) p_e\left(s_i = \omega_m\right) p_e\left(s_{i-}^{(n)}\right)}{\sum_{\omega_k \in \Omega} \sum_{n=1}^{N} p\left(r \mid s_i = \omega_k, s_{i-}^{(n)}\right) p_e\left(s_i = \omega_k\right) p_e\left(s_{i-}^{(n)}\right)}, \tag{2.33}$$

where $\Omega = \{\omega_1, \ldots, \omega_P\}$ denotes the set of symbols for all possible values of s_i.

When $\Omega = \{-1, +1\}$, $p(s_i|r, \lambda_e)$ in the form of the log-likelihood ratio can be expressed as

$$\begin{aligned} \mathrm{LLR}\left(s_i \mid r\right) &= \ln \frac{\sum_{n=1}^{N} p\left(r \mid s_i = 1, s_{i-}^{(n)}\right) p_e\left(s_{i-}^{(n)}\right)}{\sum_{n=1}^{N} p\left(r \mid s_i = -1, s_{i-}^{(n)}\right) p_e\left(s_{i-}^{(n)}\right)} + \ln \frac{p_e\left(s_i = 1\right)}{p_e\left(s_i = -1\right)} \\ &= \mathop{\mathrm{LLR}}_{\mathrm{ext}}\left(s_i \mid r\right) + \mathop{\mathrm{LLR}}_{\mathrm{prior}}\left(s_i \mid r\right), \end{aligned} \tag{2.34}$$

where $p\left(r \mid s_i, s_{i-}^{(n)}\right) = \frac{1}{\left(\pi\sigma_n^2\right)^{N_r}} \exp\left(-\frac{1}{\sigma_n^2}\|\text{r}-\text{Hs}\|^2\right)$. According to the posterior probability of each symbol, the posterior probability of each coded bit can be computed as

$$p(c_{ik} = 1|r) = \sum_{\omega_m; c_{ik}=1} p(s_i = \omega_m|r), \tag{2.35}$$

where $s_i = \text{map}(c_i)$ and $\text{map}(\cdot)$ denotes the mapping relationship between encoded bits and symbols. In addition, $c_i = (c_{i1}, \dots, c_{iK})$ and $K = \log_2 X$, where X means the constellation diagram size.

The specific steps of Gibbs Sampler are shown in Algorithm 2.4.

Algorithm 2.4 Gibbs Sampler

1: **Initialize** An initial sampling sequence $\mathbf{s}^{(-N_b)}$ is generated according to $p_e(s_i)$;
2: **for** $n = -N_b + 1$ to N_e **do**
3: Obtain the sampling point $s_1^{(n)}$ from $p(s_1|s_2^{(n-1)}, s_3^{(n-1)}, \dots, s_M^{(n-1)}, r)$;
4: Obtain the sampling point $s_2^{(n)}$ from $p(s_2|s_1^{(n)}, s_3^{(n-1)}, \dots, s_M^{(n-1)}, r)$;
5: …
6: Obtain the sampling point $s_M^{(n)}$ from $p(s_M|s_1^{(n)}, s_3^{(n)}, \dots, s_{M-1}^{(n)}, r)$;
7: **end for**
8: **Output** $\mathbf{s}^{(N_e)}$

Example 2.4 Consider a 2×2 MIMO system with 4-QAM modulation method. Assume that the initial signal is a 100-byte signal vector $\mathbf{s}$ with the corresponding value space $\Omega = \{1 \pm i, -1 \pm i\}$, and the signal passes through a randomly generated Rayleigh channel $\mathbf{H} \sim \mathcal{CN}(\mathbf{0}, \sqrt{2}\mathbf{I})$ with system noise $\mathbf{n} \sim \mathcal{CN}(\mathbf{0}, \frac{1}{\sqrt{\text{SNR}}}\mathbf{I})$. Assume that the number of iterations is 100 and one Markov chain is used for a single iteration.

To conduct a numerical analysis of the MCMC detector's performance, the simulation of this detection method is provided. To start with, a 100-byte signal vector $\mathbf{s}_0$ is generated as the initial detection signal. Subsequently, the acceptance probability $P_{\omega_{m,j}}$ of the jth bit signal taking the value ω_m is calculated. Then, if the acceptance probability satisfies the demands, the jth bit signal is updated to the corresponding variable ω, otherwise it is unchanged. This process is then repeated until all iterations have been completed.

The comparison between ZF detector and MCMC detector in 4-QAM 2×2 MIMO system is shown in Fig. 2.13. From the figure, it can be observed that the BER of MCMC is higher than that of ZF at the beginning. However, as the number of iterations increases, the BER of MCMC decreases. After about 50 iterations, the

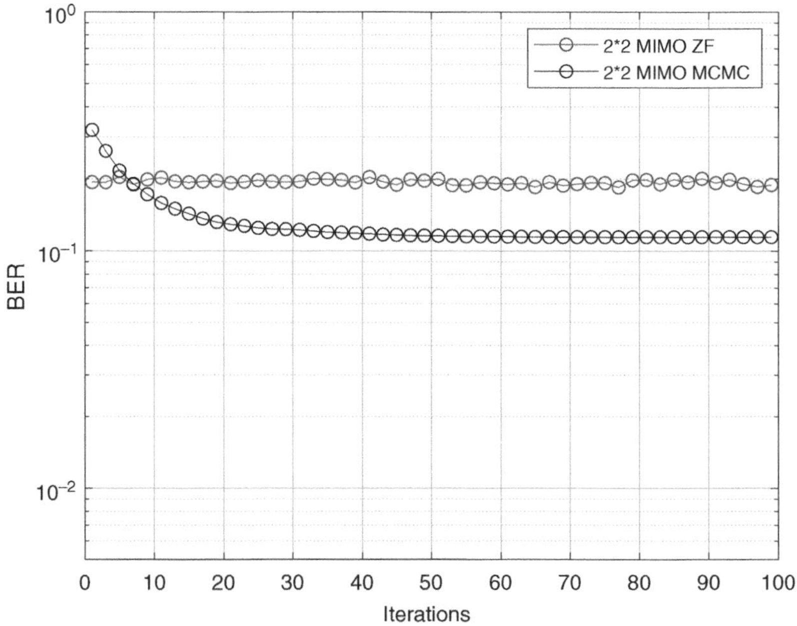

Figure 2.13 Comparison between ZF detector and MCMC detector when SNR = 5 dB.

BER of MCMC remains stable, which illustrates that MCMC detector has better performance than ZF detector.

Figure 2.14 shows that the BER of MCMC detector in 4-QAM 2×2 MIMO system decreases stably with SNR from 0 to 20 dB, which reveals that the performance of MCMC detector can be improved under the increased SNR. There have different approaches to improve the performance of MCMC detector, which will be introduced in Chapter 5.

2.2.6 VI Detection

In the MIMO detection problem, in order to detect the signal vector $\mathbf{s}$ from the received signal vector $\mathbf{r}$, it may be simpler to obtain an approximation of the distribution of $\mathbf{s}$ using the VI detection method.

Generally, for those unknown distribution of some random variables, sometimes their probability density functions (pdf) $p(\mathbf{s})$ are intractable to compute. In this case, the basic idea of the VI method is to construct a distribution $q(\mathbf{s})$, namely the variational distribution, which can be easily computed, and continuously decrease the error between the variational distribution and the actual distribution of $\mathbf{s}$ until the variational distribution is convergent. Once the convergence is achieved, $q(\mathbf{s})$ can be regarded as an approximation of $p(\mathbf{s})$.

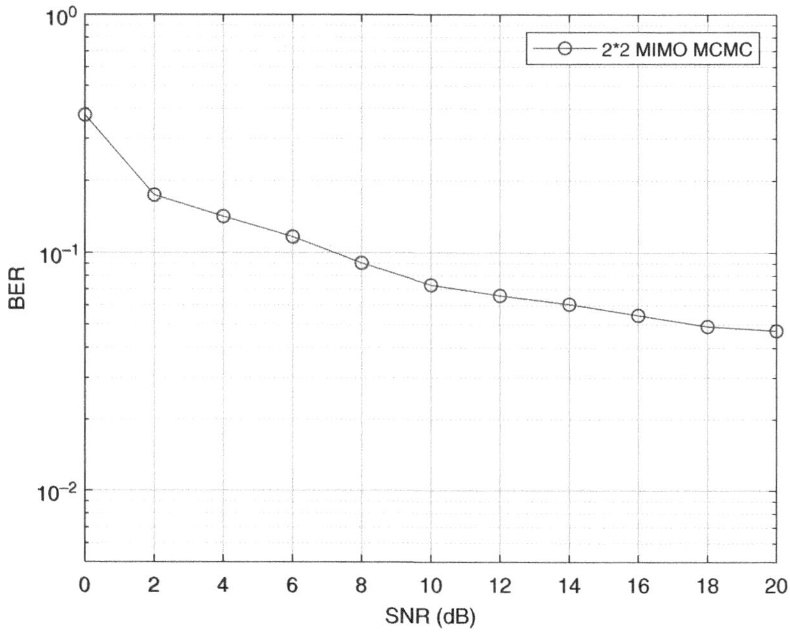

Figure 2.14 BER of MCMC detector in 4-QAM 2×2 MIMO system.

The algorithm of VI detection mainly consists of two parts: assumption of variational distribution and computation of Kullback–Leibler (KL) divergence. The first procedure is the assumption of variational distribution. Since the inference of unknown variable can be regarded as the problem of computing the posterior distribution of that variable, the detection of signal vector **s** can be replaced by the computation of the distribution of **s**. Additionally, sometimes it is challenging to calculate the distribution of **s**. In this case, the VI detection can provide a method to suppose a variational distribution to fit the distribution of **s**, which can simplify the computation. To select one family for the appropriate distribution, one effective type is the mean-field variational family, which supposes that the elements of **s** are independent of each other. Suppose the variational distribution in the mean-field approximation of each s_k is $q_k(s_k)$, then $q(s)$ can be expressed as [25]

$$q(s) = \prod_k q_k(s_k). \tag{2.36}$$

The second procedure is the computation of KL divergence. Generally, the error and distance between the designed variational distribution of **s** and its original distribution are supposed to be decreased to improve the similarity between these two distributions, so that the accuracy of variational distribution can be improved.

In this case, the KL divergence, defined as the difference between the information entropy of variational distribution and that of the original distribution, represents the comparability between the variational distribution of **s** and its initial distribution. If one of two distributions is the actual distribution, and the other is the fitted distribution, then the KL divergence is equal to the difference between the cross entropy and information entropy of the actual distribution as

$$\mathrm{KL}(p(\mathbf{s})||q(\mathbf{s})) = \sum_{i=1}^{K} [p(s_i)\log p(s_i) - p(s_i)\log q(s_i)]. \tag{2.37}$$

Under a lower value of KL divergence, the variational distribution of **s** can fix and approximate its original distribution better.

For instance, if the KL divergence is defined as $\mathrm{KL}(q(\mathbf{s})||\Pr(\mathbf{s}|\mathbf{r}))$, where $\Pr(\mathbf{s}|\mathbf{r})$ represents the initial distribution of **s**, then the problem is to minimize the KL divergence, as well as to consider the optimization problem expressed as

$$q^*(\mathbf{s}) = \underset{q(\mathbf{s})\in\Omega}{\operatorname{argmin}}\, \mathrm{KL}(q(\mathbf{s})||\Pr(\mathbf{s}|\mathbf{r})), \tag{2.38}$$

where Ω denotes the collection of all possible distributions of variable s. The KL divergence can be computed as [26]

$$\begin{aligned}
\mathrm{KL}(q(\mathbf{s})||\Pr(\mathbf{s}|\mathbf{r})) &= -\int_s q(\mathbf{s})\log\left(\frac{\Pr(\mathbf{s}|\mathbf{r})}{q(\mathbf{s})}\right)ds \\
&= \int_s q(\mathbf{s})\log q(\mathbf{s})ds - \int_s q(\mathbf{s})\log \Pr(\mathbf{s}|\mathbf{r})ds \\
&= \mathbb{E}_q[\log q(\mathbf{s})] - \mathbb{E}_q[\log \Pr(\mathbf{s}|\mathbf{r})] \\
&= \mathbb{E}_q[\log q(\mathbf{s})] - \mathbb{E}_q\left[\log\left(\frac{p(\mathbf{s},\mathbf{r})}{p(\mathbf{r})}\right)\right] \\
&= \mathbb{E}_q[\log q(\mathbf{s})] - \mathbb{E}_q[\log p(\mathbf{s},\mathbf{r})] + \mathbb{E}_q[\log p(\mathbf{r})].
\end{aligned} \tag{2.39}$$

Since there are no relationships between $\log p(\mathbf{r})$ and $q(\mathbf{s})$, then $\mathbb{E}_q[\log p(\mathbf{r})] = \log p(\mathbf{r})$. Therefore, Eq. (2.39) can be computed as

$$\mathrm{KL}(q(\mathbf{s})||\Pr(\mathbf{s}|\mathbf{r})) = \mathbb{E}_q[\log q(\mathbf{s})] - \mathbb{E}_q[\log p(\mathbf{s},\mathbf{r})] + \log p(\mathbf{r}). \tag{2.40}$$

If $q(s)$ is regarded as a variable, then $\log p(r)$ is a constant. Therefore, the aforementioned optimization problem to minimize the value of KL divergence is equivalent to the problem to maximize the value of evidence lower bound (ELBO), which is expressed as

$$\mathrm{ELBO}(q(\mathbf{s})) = \mathbb{E}_q[\log p(\mathbf{s},\mathbf{r})] - \mathbb{E}_q[\log q(\mathbf{s})]. \tag{2.41}$$

Thus, the final aim is to find out the optimal solution as

$$q^*(\mathbf{s}) = \underset{q(\mathbf{s})\in\Omega}{\operatorname{argmax}}\{\mathbb{E}_q[\log p(\mathbf{s},\mathbf{r})] - \mathbb{E}_q[\log q(\mathbf{s})]\}. \tag{2.42}$$

With the maximum ELBO and the minimum KL divergence, the variational distribution of $\mathbf{s}$ can fix the original distribution with less error.

The specific steps of VI detection are shown in Algorithm 2.5.

Algorithm 2.5 VI Detection

Input: $\mathbf{r}$, $\mathbf{H}$, $\mathbf{s}$, $\mathbf{n}$

Output: variational distribution of $\mathbf{s}$

1: Assume the variational distribution of $\mathbf{s}$, denoted by $q(\mathbf{s})$, based on Eq. (2.36);
2: Compute the ELBO based on Eq. (2.41);
3: Find the variational distribution $q^*(\mathbf{s})$ that maximizes the ELBO, based on Eq. (2.42);
4: Return $q^*(\mathbf{s})$ as the final decided variational distribution of $\mathbf{s}$;

Example 2.5 Consider a 2×2 MIMO system and the channel is an additive white Gaussian noise (AWGN) channel. Let $\mathbf{s}$ be the transmit signal vector with $\mathbf{s} = [s_1\ s_2]$, where s_1 and s_2 represent the transmitted binary bit, i.e., 0 or 1, of the first and the second transmit antenna, respectively. The channel matrix $\mathbf{H}$ is generated using random complex Gaussian distribution and the noise vector is assumed to be $\mathbf{n} \sim \mathcal{CN}(\mathbf{0}, N_0\mathbf{I})$. The probability that $s_k = 1\,(k = 1, 2)$ is $p(s_k = 1) = 0.025$. The symbol energy E_s is set to 2.

For each experiment, the main process is to compute ELBO in each iteration and compare the difference between each two consecutive values of ELBO to obtain the optimal value of ELBO. In this case, the accuracy ϵ is set to 1×10^{-6}. If the difference between two consecutive values of ELBO is lower than ϵ, then the algorithm will be stopped to obtain the optimal values of ELBO, and the final result of variational distribution of $\mathbf{s}$ can be computed as well.

Figure 2.15 shows the result of BER with the change of SNR in dB, where SNR is varied from 0 to 20 dB. It can be obtained that BER decreases with the increased SNR. This means that with SNR increases, the outcome that using the variational distribution of $\mathbf{s}$ to fix its real distribution becomes more accurate, and the variational distribution of $\mathbf{s}$ can represent its true distribution more effectively. Thus, with a relatively high SNR, the VI detection can provide a strategy to estimate the real distribution of the target signal $\mathbf{s}$.

2.2.7 CS Detection

The CS theory was first proposed by Donoho [27]. As a new sampling theory breaking Nyquist's sampling theory, it has been widely concerned by researchers. Based on the sparsity of signals, CS theory requires that signals can be sparsely

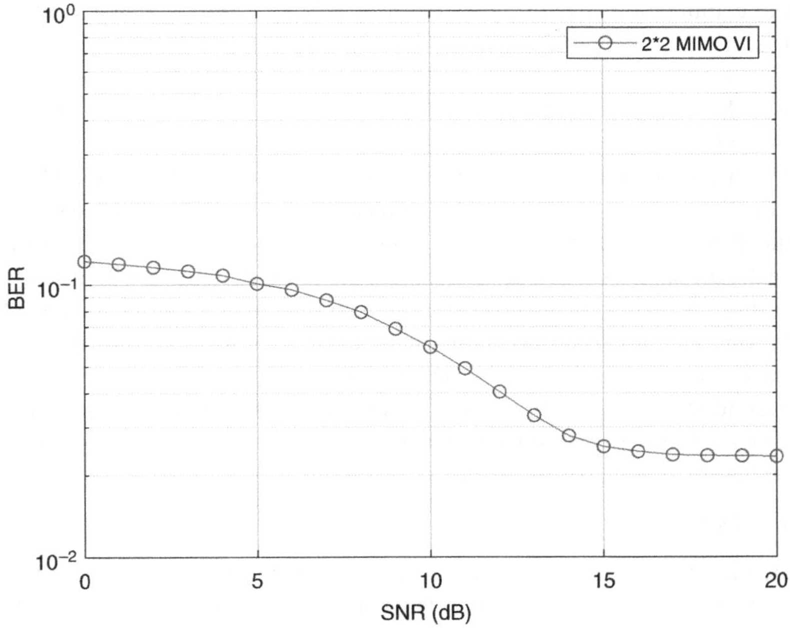

Figure 2.15 BER of VI detection in 2 × 2 MIMO system.

expressed in a certain variation domain. In CS theory, the sampling rate of the signal, which is no longer determined by the highest frequency of the signal, is closely related to the information structure of the signal. Compared with Nyquist's theory, CS theory usually reconstructs the original sparse signal from a small number of observed values. Due to these characteristics, CS can be used for signal detection in MIMO systems. Besides, in some IoT scenarios, the device activation is usually sporadic and only a small number of devices are active in each time slot, and it is more effective to utilize the CS algorithm in IoT systems than the traditional algorithms.

CS detection mainly includes three aspects: signal sparse representation, compression measurement, and signal reconstruction, which are shown in Fig. 2.16.

The first step is the signal sparse representation. Assume that $\mathbf{s}$ is a column vector of dimension $M \times 1$ in $\mathbb{R}^M$ space, and any signal in $\mathbb{R}^M$ space can be written

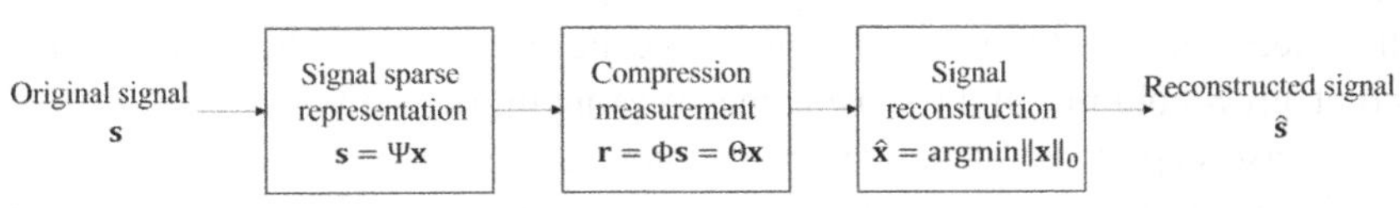

Figure 2.16 The process of compressive sensing.

as a linear combination of an orthonormal basis $\{\Psi_i\}_{i=l}^{M}$. Then, the signal $\mathbf{s}$ can be written as

$$\mathbf{s} = \sum_{i=1}^{M} x_i \Psi_i = \Psi \mathbf{x}, \quad i = 1, 2, \ldots, M, \tag{2.43}$$

where Ψ is the matrix formed by the orthonormal basis $\{\Psi_i\}_{i=1}^{M}$, and $\mathbf{s}$ is the $M \times 1$ dimensional column vector generated by projecting the coefficients of the signal $\mathbf{x}$ into the Ψ domain. If there are at most K nonzero elements in $\mathbf{s}$ and K is much less than M, the signal $\mathbf{x}$ is sparse or compressible in the Ψ domain with sparsity K. Note that CS theory is based on the premise that signals can be expressed sparsely. Some common transform bases include Fourier transform basis, discrete cosine basis, wavelet transform basis, Gabor basis, etc.

The second step is the compressive measurement. The receiver projects the received signal from the high-dimensional space to the low-dimensional space by using the observation matrix Φ, which can be shown as

$$\mathbf{r} = \Phi \mathbf{s} = \Phi \Psi \mathbf{x} = \Theta \mathbf{x}. \tag{2.44}$$

Here, the observation matrix Φ is a flat matrix of dimension $N \times M$ and $\mathbf{r}$ is a column observation vector of length N. The matrix $\Theta = \Phi\Psi$ of dimension $N \times M$ is called the sensing matrix, where the sensing matrix Θ computes a sparse vector of length M to an observation vector $\mathbf{r}$ of length N. Since $N < M$, it maps the signal from high-dimensional space to low-dimensional space. To losslessly reconstruct $\mathbf{s}$ or $\mathbf{x}$ from $\mathbf{r}$, $\mathbf{r}$ must contain all useful information.

It is also pointed out that the sensing matrix Θ should satisfy the RIP to ensure that the CS algorithm can accurately reconstruct the original sparse signal. The RIP criterion states that, for any K sparse signal $\mathbf{x}$, if there exists a constant $\delta \in (0, 1)$ such that the sensing matrix Θ can satisfy

$$(1 - \delta)\|\mathbf{x}\|_2^2 \leq \|\Theta \mathbf{x}\|_2^2 \leq (1 + \delta)\|\mathbf{x}\|_2^2, \tag{2.45}$$

then the sensing matrix Θ is said to satisfy the RIP property, and δ is called the constrained isometry constant. The RIP criterion ensures that the mapping has uniqueness. That is, the sensing matrix Θ will not map various sparse signals into the same observation vector. Most uniformly distributed random matrices can satisfy the RIP property well. Common sensing matrices include Bernoulli random matrices, partial Fourier matrices, independent and equally distributed Gaussian random matrices, Rademacher matrices, and partial Hadamard matrices. Generally, a necessary condition for CS to guarantee the uniqueness of the reconstruction is that any K columns of the sensing matrix Θ are linearly independent.

The last step is the signal reconstruction. If the sensing matrix Θ satisfies certain conditions, then the compressed low-dimensional vector can reconstruct the original high-dimensional signal. The process of solving the original high-dimensional signal is called the signal reconstruction. When the signal is sparsely expressed, since the orthogonal basis matrix $\mathbf{\Phi}$ is known, the reconstruction of the original signal $\mathbf{s}$ and the sparse vector $\mathbf{x}$ are equivalent. Equation (2.44) is a linear equation and solving the unknown vector $\mathbf{x}$ from the linear equation is originally a linear regression problem. However, because $N \ll M$, Eq. (2.44) is an underdetermined equation. Since CS restricts the sparsity of the vector $\mathbf{s}$, the multi-solution problem is further limited and the CS algorithm can use the sparsest solution to estimate the original signal, so that Eq. (2.44) is eventually evolved into solving the optimization problem with sparse constraints as

$$\begin{aligned} &\hat{\mathbf{x}} = \text{argmin}\ \|\mathbf{x}\|_0\ , \\ &\text{s.t.}\ \mathbf{r} = \mathbf{\Theta}\mathbf{x}. \end{aligned} \tag{2.46}$$

However, the optimization problem based on the L_0 minimum norm constraint is a non-deterministic polynomial (NP) hard problem. Solving Eq. (2.46) requires considering all possible linear combinations of the positions of the non-zero coefficients in $\mathbf{x}$, which is, $C\begin{pmatrix} M \\ K \end{pmatrix}$. As it is troublesome to directly solve Eq. (2.46), scholars generally convert it into the problem of finding a sub-optimal solution, where the methods mainly include greedy algorithm, convex relaxation algorithm, and statistical optimization algorithm.

The specific steps of CS detection are shown in Algorithm 2.6.

Algorithm 2.6 CS Detection

Input: $\mathbf{r}$, $\mathbf{\Psi}$

Output: reconstructed signal $\hat{\mathbf{s}}$

1: Construct measurement matrix $\mathbf{\Phi}$;
2: Perform compression measurement on the received signal $\mathbf{r}$ based on Eq. (2.44);
3: Choose a reconstruction algorithm, such as greedy algorithm or convex optimization algorithm;
4: Obtain the reconstructed signal $\hat{\mathbf{s}}$ based on (2.46).

Example 2.6 Suppose a 256-bit signal vector $\mathbf{s}$ has the sparsity of 50, i.e., there are 50 bits with a value of 1. Assume that the length of low-dimensional observation vector $\mathbf{r}$ is varied from 64 to 128. An identity matrix is used to be sparse matrix, i.e., $\mathbf{\Psi} = \mathbf{I}(256)$, while $\mathbf{I}(L)$ represents the identity matrix of order L. The measurement matrix $\mathbf{\Phi}$ is assumed to be a Gaussian matrix. Then the observation vector $\mathbf{r}$ can be obtained according to Eq. (2.44). We consider the basis pursuit (BP) algorithm, one of the classical reconstruction algorithms in CS, to obtain the reconstructed signal $\hat{\mathbf{s}}$. The reconstruction residual is shown in Fig. 2.17.

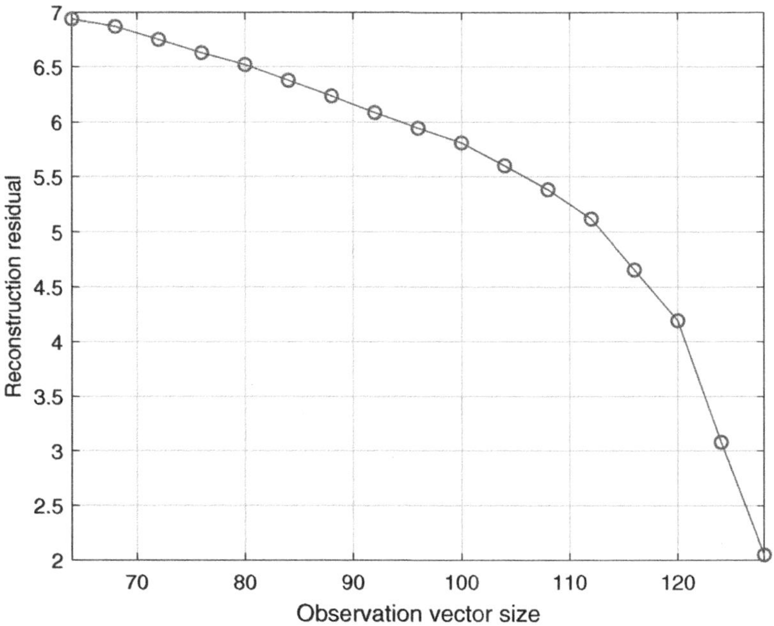

Figure 2.17 The reconstruction residual with different observation vector sizes.

As can be seen from Fig. 2.17, the larger the size of the observed vector, the smaller the reconstruction residual. That is, the smaller the gap between the reconstructed vector and the original vector. This shows that although CS algorithm can reduce the size of the transmitted data, in order to ensure the accuracy of recovery, the compressed vector should not be too small.

2.3 Conclusion and Remarks

In this chapter, the background of IoT signal detection is introduced, including RA strategies, traditional signal detection methods, and some new signal detection algorithms. Among them, RA strategies can be divided into grant-based RA and grant-free RA, which are distinguished by whether to obtain the reserved channel information. Except for some traditional signal detection methods such as ML, ZF, MMSE, and some popular signal detection methods such as MCMC detection, VI detection, CS detection, are discussed in detail as well. In Chapters 3–6, we will apply these methods into IoT signal detection.

3

Sparse Signal Detection for Multiple Access

- As the number of Internet of Things (IoT) devices grows explosively in wireless networks, the next generation multiple access (NGMA) schemes are considered to support massive access for a large number of devices. Since multiple-input multiple-output (MIMO) has become one of the essential techniques for providing a sufficient degree of spatial freedom and improving the spectral efficiency, there have been various transmission schemes over MIMO channels to achieve the tradeoff between diversity and multiplexing gains, in which spatial modulation (SM) [28] is a newly proposed technique to select among multiple transmit antennas to obtain high data rates by introducing spatial domain. Considering that the signal vector of SM systems contains a number of zero elements, the information vector being transmitted becomes sparse. Thus, compressive sensing (CS) reconstruction algorithms [29] can be applied to SM systems to detect sparse signals, which enable the reconstruction of compressible or sparse signals. In this chapter, we introduce some sparse signal detection approaches for multiple access of IoT devices [30].

3.1 System Model

Consider a MIMO system with M inputs and N outputs. The received signal at a receiver, denoted by $\mathbf{y} \in \mathbb{C}^{N\times 1}$, is given by

$$\mathbf{y} = \mathbf{C}\mathbf{b} + \mathbf{n}, \tag{3.1}$$

where $\mathbf{C}$ is an $N \times M$ matrix, $\mathbf{b} \in \mathbb{C}^{M\times 1}$ is the signal to be detected, and $\mathbf{n} \sim \mathcal{CN}(\mathbf{0}, N_0\mathbf{I})$ is the background noise vector. Let $\mathbf{b} = [b_1 \dots b_M]^{\mathrm{T}}$, where $b_m \in \mathcal{B}$ is a discrete random variable. Here, $\mathcal{B}$ is the signal constellation. Denote by $\Pr(b_m)$ the a priori probability of b_m and assume that the b_m's are independent with each other. Denote by $f(\mathbf{y} \mid \mathbf{b})$ the likelihood function of $\mathbf{b}$. To estimate $\mathbf{b}$,

IoT Signal Detection, First Edition. Rui Han, Jingjing Wang, Lin Bai, and Jianwei Liu.

we consider either the maximum likelihood (ML) or maximum a posteriori probability (MAP) problem as

$$\begin{aligned}\hat{\mathbf{b}}_{\rm ml} &= \underset{\mathbf{b}\in\mathcal{B}^M}{\arg\max} f(\mathbf{y}|\mathbf{b}),\\ \hat{\mathbf{b}}_{\rm map} &= \underset{\mathbf{b}\in\mathcal{B}^M}{\arg\max}\ \Pr(\mathbf{b}|\mathbf{y}),\end{aligned} \tag{3.2}$$

respectively, where $\Pr(\mathbf{b}\,|\,\mathbf{y}) \propto f(\mathbf{y}\,|\,\mathbf{b})$ represents the a posteriori probability of $\mathbf{b}$. The problem in Eq. (3.2) is referred to as a MIMO detection problem. In general, it is challenging to find the ML or MAP solution due to prohibitively high computational complexity. Thus, a number of low-complexity approaches are investigated to tackle the issues [31, 32], e.g., zero forcing (ZF) detection and minimum mean square error (MMSE) detection. However, those sub-optimal approaches require that $M \leq N$ in general.

In MIMO communications with multiple antennas, the signal vector $\mathbf{b}$ becomes sparse in SM or generalized SM, where a transmitter equipped with M antennas can transmit signals by activating one or few antennas at a time [33, 34]. In this case, the signal constellation is to be extended to $\tilde{\mathcal{B}} = \mathcal{B} \cup \{0\}$ in order to take into account the case that no signals are transmitted from a transmit antenna. As for code division multiple access (CDMA)-based random access (RA) with M nodes where each column of $\mathbf{C}$ is a dedicated signature sequence for a node [35–37]. Let $\mathbf{c}_m$ and b_m be the signature sequence and the signal of the mth node, respectively. Since a few nodes have signals to transmit at a time, $\mathbf{b}$ becomes sparse. Note that according to Eq. (3.1), $\mathbf{c}_m$ is not only the signature sequence but also the spreading sequence for node m as in CDMA.

Let Q be the sparsity of $\mathbf{b}$, i.e., the number of non-zero elements is Q or $||\mathbf{b}||_0 = Q$. For simplicity, let $\tilde{\mathcal{B}} = \{0, 1\}$. In this case, the number of the possible signal vectors for $\mathbf{b}$ is $\binom{M}{Q}$. If Q is sufficiently small (e.g., $Q \leq 2$), since $\binom{M}{Q} \approx \frac{M^Q}{Q!}$, the ML or MAP detection may be carried out using an exhaustive search. However, in general, if M^Q is not small, it is troublesome to perform the ML or MAP detection due to the high computational complexity.

Since $Q \ll M$, $\mathbf{b}$ becomes sparse and CS-based approaches can be used as in SM [38] for the signal detection and in [37, 39] for RA. In particular, $\mathbf{C}$ in Eq. (3.1) becomes the measurement matrix (that has more columns than rows) and $\mathbf{b}$ is a sparse vector. A low-complexity CS algorithm [40] can be used to find $\mathbf{b}$ from $\mathbf{y}$ by exploiting the sparsity of $\mathbf{b}$. However, although the complexity of CS-based approaches is lower than that of ML or MAP, it is still proportional to M. Thus, for a large M, it would be desirable to decrease the complexity.

For NGMA in IoT, we assume that a base station (BS) is equipped with a number of antennas, saying N antennas. In addition, there are a large number of users, saying G users, where each user is equipped with multiple antennas, saying B antennas. Then, there are $M = BG$ inputs. Each user can be active or inactive, and

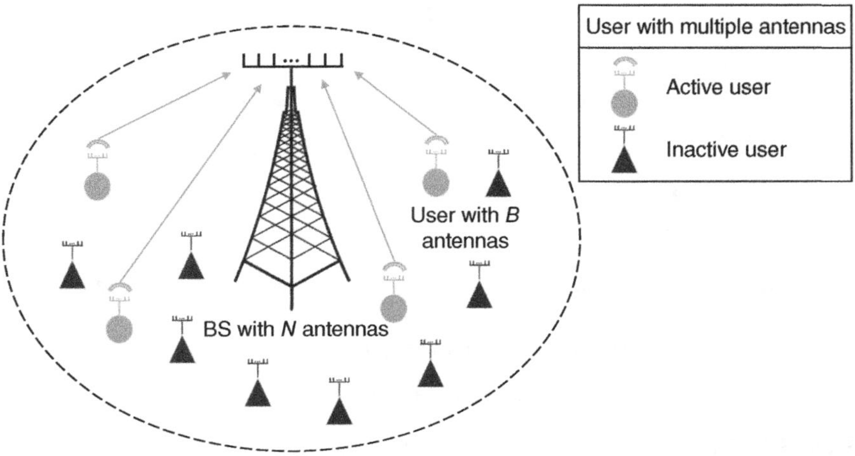

Figure 3.1 Structure of NGMA systems.

an active user uses SM to transmit his/her information as shown in Fig. 3.1. In general, it is expected that $M \gg N$, while only a fraction of M inputs (saying Q inputs) are not zero (thanks to the sparse user activity and SM), which makes the input vector sparse. CDMA-based NGMA can also be considered, where each user has B spreading codes of length N for SM. With multiple antennas and spreading codes, a large system for NGMA can also be formed. As a result, in NGMA, it is necessary to deal with a large MIMO matrix $\mathbf{C}$, and a low-complexity algorithm to solve a large MIMO detection problem by effectively exploiting the sparsity is highly desirable.

3.2 Sparse Signal Detection

In this section, we introduce a tree search-based approach to detect sparse signal by exploiting the sparsity with $Q \ll M$.

3.2.1 Tree Search-based Approach

Firstly, we introduce the method of tree search with a binary tree. At the root node, the presence of any active signals is to be decided with $\mathbf{y}$ as shown in Fig. 3.2. If any signal is presented in $\mathbf{y}$, the signal index set, $\{1, \dots, M\}$, is divided into two subsets, namely $\{1, \dots, \lfloor M/2 \rfloor\}$ and $\{\lfloor M/2 \rfloor + 1, \dots, M\}$. Then, the decision on the absence or presence is carried out with the subsets. The same process is repeated with smaller subsets.

As shown in Fig. 3.2, there can be four subsets. If any of them do not have any signals, they can be excluded. For example, if the third and fourth subsets do not have

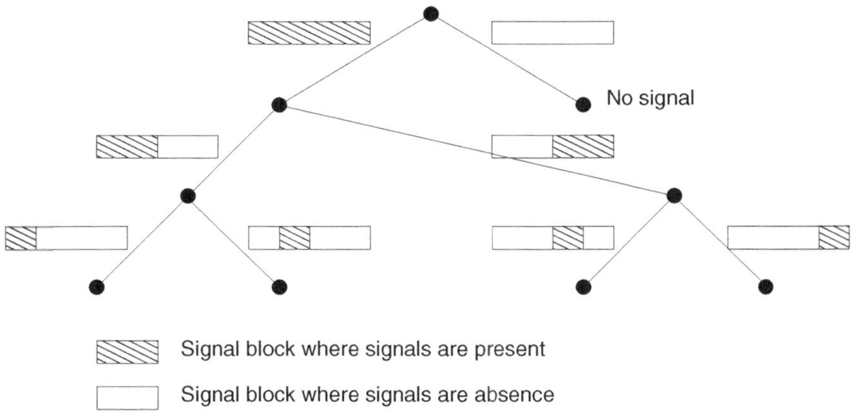

Figure 3.2 The tree search for the presence of (sparse) signals in a block.

any signals, the signal detection can be run with the submatrix of $\mathbf{C}$, $[\mathbf{c}_1 \dots \mathbf{c}_{\lfloor M/2 \rfloor}]$. Thus, the complexity can be reduced. Note that, an n-ary tree can also be considered, where $n \geq 2$.

Motivated by tree search, we propose a two-stage (TS) approach for sparse signal detection, where TSs are summarized as follows:

- **Stage 1**: Active variable detection using coordinate ascent variational inference (CAVI) algorithm to decide active groups or inactive groups.
- **Stage 2**: CS-based algorithms (e.g., the orthogonal matching pursuit (OMP) algorithm, the least absolute shrinkage and selection operator (LASSO) algorithm, and the approximate message passing (AMP) algorithm) are carried out with the signals belonging to active groups, while the signals associated with inactive groups are removed.

Then, we consider a special case with TSs with details. We assume that M columns are divided into K groups and the aggregated signal for group k becomes

$$\mathbf{g}_k = \sum_{m \in \mathcal{I}_k} \mathbf{c}_m b_m, \tag{3.3}$$

where $\mathcal{I}_k$ is the index set of the columns for group k. For convenience, let $D = \frac{M}{K} = |\mathcal{I}_k|$ for all k, where D is a positive integer. It is assumed that $\cup_{k=1}^{K} \mathcal{I}_k = \{1, \dots, M\}$ and $\mathcal{I}_{k'} \cap \mathcal{I}_k = \emptyset$, $k \neq k'$. In order to detect the activity of group k, we define the activity variable of group k as

$$x_k = \begin{cases} 0, & \text{if } b_m = 0 \text{ for all } m \in \mathcal{I}_k \,, \\ 1, & \text{o.w.} \end{cases} \tag{3.4}$$

It is indicated that there are no signals in group k if $x_k = 0$, and vice versa. In addition, we assume that b_m is the independently identically distribution with

$\Pr(b_m \neq 0) = p$ and $\Pr(b_m = 0) = 1 - p$, where p is the activity probability. The (conditional) signal variance when b_m is active, i.e., $b_m \neq 0$, can be given by

$$\mathbb{E}[|b_m|^2 \mid b_m \neq 0] = E_s, \tag{3.5}$$

where E_s represents the symbol energy for $b_m \in \mathcal{B}$, with $\mathbb{E}[b_m \mid b_m \neq 0] = 0$.

Clearly, if $x_k = 0$, we have $\mathbf{g}_k = \mathbf{0}$. On the other hand, if $x_k = 1$, under the Gaussian approximation, we assume that

$$\mathbf{g}_k \sim \mathcal{CN}(\mathbf{0}, \mathbf{R}_k), \tag{3.6}$$

where it is assumed that $\mathbb{E}[\mathbf{g}_k \mid x_k = 1] = \mathbf{0}$ and

$$\mathbf{R}_k = \mathbb{E}[\mathbf{g}_k \mathbf{g}_k^{\mathrm{H}} \mid x_k = 1]. \tag{3.7}$$

It can be shown that

$$\begin{aligned}\mathbf{R}_k &= \sum_{m \in \mathcal{I}_k} \sum_{m' \in \mathcal{I}_k} \mathbb{E}[\mathbf{c}_m \mathbf{c}_{m'}^{\mathrm{H}}] \mathbb{E}[b_m b_{m'} \mid x_k = 1] \\ &= \sum_{m \in \mathcal{I}_k} \mathbf{c}_m \mathbf{c}_m^{\mathrm{H}} \mathbb{E}[|b_m|^2 \mid x_k = 1] \\ &= \mathbf{C}_k \mathbf{C}_k^{\mathrm{H}} \beta_k,\end{aligned} \tag{3.8}$$

where $\mathbf{C}_k$ is a matrix consisting of the columns of $\mathbf{C}$ corresponding to the index set $\mathcal{I}_k$ and

$$\beta_k = \mathbb{E}[|b_m|^2 \mid x_k = 1] = \frac{E_s p}{1 - (1-p)^D}. \tag{3.9}$$

Note that the probability that $x_k = 0$ is $(1-p)^D$ and $x_k = 1$ is $1 - (1-p)^D$. In addition, since β_k is same for all k, let $\beta_k = \beta = \frac{E_s p}{1-(1-p)^D}$. Consequently, if we assume that

$$\mathbf{a}_k \sim \mathcal{CN}\left(\mathbf{0}, \mathbf{R}_k\right), \tag{3.10}$$

then the received signal $\mathbf{y}$ can be re-written as

$$\mathbf{y} = \sum_{k=1}^{K} \mathbf{a}_k x_k + \mathbf{n}, \quad x_k \in \{0, 1\}, \tag{3.11}$$

with the aggregated signals, $\{\mathbf{a}_k\}$, and the activity variables, $\{x_k\}$.

As discussed above, the TS approach can be shown as follows:

- In the first stage, with Eq. (3.11), we decide active and inactive groups with the activity variables, $\{x_k\}$.
- In the second stage, the signals associated with inactive groups are removed, and the sparse signal detection is carried out with the signals belonging to active groups.

Thus, the computational complexity of the second stage can be lower if there are more inactive groups.

Example 3.1 Let $M = 8$, $N = 3$, and $K = 4$ with $D = 2$. In addition, let $\mathbf{b} = [0\,0\,0\,1\,0\,0\,1\,0]^{\mathrm{T}}$. Then, $x_1 = x_3 = 0$, while $x_2 = x_4 = 1$. If all the x_k's are correctly detected, in Stage 2, we have

$$\mathbf{y} = [\mathbf{C}_2\,\mathbf{C}_4][b_3\,b_4\,b_7\,b_8]^{\mathrm{T}} + \mathbf{n}.$$

In this example, $[\mathbf{C}_2\,\mathbf{C}_4]$ becomes a 3×4 matrix. If we consider a conventional approach without TSs, $\mathbf{C}$ is a 3×8 matrix. Thus, compared with the conventional single-stage (SS) approach, the complexity of the TS approach can be lower and the performance can be improved as well (as the number of columns of the measurement matrix is smaller), provided that the x_k's are correctly detected in Stage 1.

3.2.2 VI Detection Algorithm

In the following, we discuss the detection of aggregated signals with the model in Eq. (3.11) under the Gaussian approximation [41].

For convenience, let

$$\mathbf{x} = [x_1\ \dots\ x_K]^{\mathrm{T}} \quad \text{and} \quad \mathbf{x}_{-k} = [x_1\ \dots\ x_{k-1}\,x_{k+1}\ \dots\ x_K]^{\mathrm{T}}.$$

To decide x_k, using Eq. (3.11), it can be shown that

$$\mathbf{y} = \mathbf{a}_k x_k + \mathbf{u}_k, \tag{3.12}$$

where $\mathbf{u}_k = \sum_{q \neq k} \mathbf{a}_q x_q + \mathbf{n}$. In addition, we have two hypotheses: $\mathcal{H}_0$ for $x_k = 0$ and $\mathcal{H}_1$ for $x_k = 1$. For each hypothesis, under the Gaussian approximation in Eq. (3.11) or (3.12), it can be shown that

$$\mathbf{y} \sim \begin{cases} \mathcal{CN}(\mathbf{0}, \mathbf{V}_k^{-1}), & \text{under } \mathcal{H}_0, \\ \mathcal{CN}(\mathbf{0}, \mathbf{V}_k^{-1} + \mathbf{R}_k), & \text{under } \mathcal{H}_1, \end{cases} \tag{3.13}$$

where $\mathbf{V}_k = (\sum_{q \neq k} \mathbf{R}_q x_q + N_0 \mathbf{I})^{-1}$. As in [42], the test statistic is given by

$$A_k = \mathbf{y}^{\mathrm{H}} \mathbf{D}_k \mathbf{y}, \tag{3.14}$$

where

$$\mathbf{D}_k = \mathbf{V}_k - (\mathbf{V}_k^{-1} + \mathbf{R}_k)^{-1}. \tag{3.15}$$

For a given threshold τ, we accept $\mathcal{H}_1$ if $A_k \geq \tau$, or $\mathcal{H}_0$ if $A_k < \tau$ according to the Neyman–Pearson test [32, 42]. However, in order to find the test statistic in Eq. (3.14), it is required to know $\mathbf{x}_{-k}$, which is not available. As a result, the hypothesis testing with A_k is considered to see an achievable performance.

Another approach is based on the MAP criterion [32], which is given by

$$\hat{\mathbf{x}} = \underset{\mathbf{x} \in \mathcal{X}}{\arg\max}\ \Pr(\mathbf{x}|\mathbf{y}), \tag{3.16}$$

where $\mathcal{X} = \{0,1\}^K$. In Eq. (3.16), since all the x_k's are jointly detected, we do not have the difficulty with A_k, i.e., we do not need $\mathbf{x}_{-k}$ in detecting x_k. However, the computational complexity can be high for a large K.

In order to find an approximate solution to Eq. (3.16), an iterative variational inference (VI) approach is presented as follows. Suppose that there is a parameter vector $\mathbf{x}$, that is to be estimated from an observation $\mathbf{y}$. In VI, instead of detecting $\mathbf{x}$, we consider the estimation of the distribution of $\mathbf{x}$ [25]. Furthermore, the mean-field approximation can be employed, where the x_k's are assumed to be independent. Let the variational distribution for each x_k be $\psi_k(x_k)$ and for the mean-field approximation, let

$$\psi(\mathbf{x}) = \prod_k \psi_k(x_k). \tag{3.17}$$

Then, we consider the following optimization problem to approximate $\Pr(\mathbf{x} \mid \mathbf{y})$ with $\psi(\mathbf{x})$ as

$$\psi^*(\mathbf{x}) = \underset{\psi(\mathbf{x}) \in \Psi}{\arg\min}\, \mathrm{KL}(\psi(\mathbf{x}) || \Pr(\mathbf{x} \mid \mathbf{y})), \tag{3.18}$$

where Ψ denotes the collection of all the possible distributions of $\mathbf{x}$ subject to the constraint in Eq. (3.17), and $\mathrm{KL}(\cdot)$ is the Kullback–Leibler (KL) divergence that is defined as

$$\mathrm{KL}(\psi(\mathbf{x}) || f(\mathbf{x})) = \sum_{\mathbf{x}} \psi(\mathbf{x}) \ln \frac{\psi(\mathbf{x})}{f(\mathbf{x})}. \tag{3.19}$$

Here, $f(\mathbf{x})$ represents the distribution of $\mathbf{x}$ with $f(\mathbf{x}) > 0$ for all $\mathbf{x} \in \tilde{\mathcal{B}}$. As shown in Eq. (3.18), we need to find $\psi(\mathbf{x})$ that is close to the a posteriori probability $\Pr(\mathbf{x} \mid \mathbf{y})$ as an approximation. Then, the minimization of KL divergence can be transformed to maximize the evidence lower bound (ELBO) [25], which becomes

$$\mathrm{ELBO}(\psi) = \mathbb{E}[\ln f(\mathbf{y}, \mathbf{x})] - \mathbb{E}[\ln \psi(\mathbf{x})]. \tag{3.20}$$

Let $\psi_{-k}(\mathbf{x}_{-k}) = \prod_{l \neq k} \psi_l(x_l)$. The CAVI algorithm [25, 43] is to update one variational distribution at a time with the other variational distribution fixed in each iteration, which can be shown as

$$\psi_k^*(x_k) \propto \exp\left(\mathbb{E}_{-k}[\ln f(x_k \mid \mathbf{x}_{-k}, \mathbf{y})]\right), \tag{3.21}$$

where $\mathbb{E}_{-k}[\cdot]$ denotes the expectation with respect to $\mathbf{x}_{-k}$ or with the distribution $\psi_{-k}(\mathbf{x}_{-k})$.

It can be shown that

$$\begin{aligned} \ln f(x_k \mid \mathbf{x}_{-k}, \mathbf{y}) &= \ln f(x_k, \mathbf{x}_{-k}, \mathbf{y}) - \ln f(\mathbf{x}_{-k}, \mathbf{y}) \\ &= \ln f(\mathbf{y} \mid x_k, \mathbf{x}_{-k}) + \ln \Pr(x_k) \\ &\quad + \ln \Pr(\mathbf{x}_{-k}) - \ln f(\mathbf{x}_{-k}, \mathbf{y}). \end{aligned} \tag{3.22}$$

Since the last two terms are independent of x_k, it can be shown that

$$\mathbb{E}_{-k}[\ln f(x_k \,|\, \mathbf{x}_{-k}, \mathbf{y})] = \mathbb{E}_{-k}[\ln f(\mathbf{y} \,|\, x_k, \mathbf{x}_{-k})] + \ln \Pr(x_k) + C, \tag{3.23}$$

where C is a constant. Let

$$\mathbf{V}(\mathbf{x}) = \left(\sum_{k=1}^{K} \mathbf{R}_k x_k + N_0 \mathbf{I} \right)^{-1},$$

and

$$\phi(\mathbf{x}) = \ln \det \left(\sum_{k=1}^{K} \mathbf{R}_k x_k + N_0 \mathbf{I} \right). \tag{3.24}$$

It can be shown that

$$f(\mathbf{y} \,|\, x_k, \mathbf{x}_{-k}) = C_0 (\det(\mathbf{V}^{-1}(\mathbf{x})))^{-1} \ \exp(-\mathbf{y}^{\mathrm{H}} \mathbf{V}(\mathbf{x}) \mathbf{y}), \tag{3.25}$$

and then,

$$\ln f(\mathbf{y} \,|\, x_k, \mathbf{x}_{-k}) = -\mathbf{y}^{\mathrm{H}} \mathbf{V}(\mathbf{x}) \mathbf{y} - \phi(\mathbf{x}) + C_1, \tag{3.26}$$

where C_0 and C_1 are constants. Let $\mathbf{V}(\mathbf{x}) = \mathbf{V}(\mathbf{x}_{-k}, x_k)$ and $\phi(\mathbf{x}) = \phi(\mathbf{x}_{-k}, x_k)$ with a slight abuse of notation. It can be shown that

$$\begin{aligned} \mathbb{E}_{-k}[\ln f(\mathbf{y} \,|\, x_k, \mathbf{x}_{-k})] &= \mathbb{E}_{-k}[-\mathbf{y}^{\mathrm{H}} \mathbf{V}(\mathbf{x}_{-k}, x_k) \mathbf{y} - \phi(\mathbf{x}_{-k}, x_k) + C_1] \\ &\approx -\mathbf{y}^{\mathrm{H}} \mathbf{V}(\mathbb{E}[\mathbf{x}_{-k}], x_k) \mathbf{y} - \phi(\mathbb{E}[\mathbf{x}_{-k}], x_k), \end{aligned} \tag{3.27}$$

where the approximation becomes tight when $\psi_k(x_k)$, $k = 1, \ldots, K$, approaches either 0 or 1. For convenience, let $\bar{x}_k = \mathbb{E}[x_k]$ and $\bar{\mathbf{x}} = [\bar{x}_1 \ \ldots \ \bar{x}_K]^{\mathrm{T}}$. We also define $\bar{\mathbf{x}}_{-k}$ as the subvector of $\bar{\mathbf{x}}$, that is obtained by removing $\bar{x}_k$.

To obtain the updating rule for $\psi_k(x_k)$ in Eq. (3.21), suppose that x_k is to be updated. Let

$$\mathbf{V}_k(\bar{\mathbf{x}}) = \left(\sum_{q \neq k} \mathbf{R}_q \bar{x}_q + N_0 \mathbf{I} \right)^{-1}. \tag{3.28}$$

Then, using the matrix inversion lemma [44], we can show that

$$\begin{aligned} \ell_k &= \mathbb{E}_{-k}[\ln f(x_k = 1 \,|\, \mathbf{x}_{-k}, \mathbf{y})] - \mathbb{E}_{-k}[\ln f(x_k = 0 \,|\, \mathbf{x}_{-k}, \mathbf{y})] \\ &\approx -\mathbf{y}^{\mathrm{H}} \left((\mathbf{V}_k^{-1}(\bar{\mathbf{x}}) + \mathbf{R}_k)^{-1} - \mathbf{V}_k(\bar{\mathbf{x}}) \right) \mathbf{y} - \phi_k(\bar{\mathbf{x}}) + \delta \\ &= -\mathbf{y}^{\mathrm{H}} \mathbf{V}_k(\bar{\mathbf{x}}) \mathbf{C}_k \left(-\frac{1}{\beta} \mathbf{I} - \mathbf{C}_k^{\mathrm{H}} \mathbf{V}_k(\bar{\mathbf{x}}) \mathbf{C}_k \right)^{-1} \mathbf{C}_k^{\mathrm{H}} \mathbf{V}_k(\bar{\mathbf{x}}) \mathbf{y} - \phi_k(\bar{\mathbf{x}}) + \delta \\ &= \mathbf{z}_k^{\mathrm{H}} \left(\frac{1}{\beta} \mathbf{I} + \mathbf{C}_k^{\mathrm{H}} \mathbf{V}_k(\bar{\mathbf{x}}) \mathbf{C}_k \right)^{-1} \mathbf{z}_k - \phi_k(\bar{\mathbf{x}}) + \delta, \end{aligned} \tag{3.29}$$

where $\mathbf{z}_k = \mathbf{C}_k^{\mathrm{H}} \mathbf{V}_k(\overline{\mathbf{x}})\mathbf{y}$, $\delta = \ln \frac{\Pr(x_k=1)}{\Pr(x_k=0)}$, and

$$\begin{aligned}\phi_k(\overline{\mathbf{x}}) &= \ln\det\left(\mathbf{V}_k^{-1}(\overline{\mathbf{x}}) + \mathbf{R}_k\right) - \ln\det\left(\mathbf{V}_k^{-1}(\overline{\mathbf{x}})\right)\\ &= \ln\det\left(\mathbf{I} + \beta\mathbf{C}_k^{\mathrm{H}}\mathbf{V}_k(\overline{\mathbf{x}})\,\mathbf{C}_k\right).\end{aligned}$$

Then, it can be shown that

$$\psi_k(x_k = 1) \approx \frac{e^{\ell_k}}{1 + e^{\ell_k}}, \tag{3.30}$$

$$\psi_k(x_k = 0) \approx \frac{1}{1 + e^{\ell_k}}. \tag{3.31}$$

Clearly, the updated value of $\overline{x}_k$ is given by $\psi_k(x_k = 1)$.

Note that, $\mathbf{V}(\overline{\mathbf{x}})$ can be updated with a new $\overline{x}_k = \psi_k(x_k = 1)$ as

$$\begin{aligned}\mathbf{V}(\overline{\mathbf{x}}_{-k}, \overline{x}_k) &= \left(\mathbf{V}_k^{-1}(\overline{\mathbf{x}}) + \overline{x}_k\mathbf{R}_k\right)^{-1}\\ &= \mathbf{V}_k(\overline{\mathbf{x}}) - \mathbf{U}_k(\overline{\mathbf{x}})\left(\frac{1}{\beta\overline{x}_k}\mathbf{I} + \mathbf{C}_k^{\mathrm{H}}\mathbf{U}_k(\overline{\mathbf{x}})\right)^{-1}\mathbf{U}_k^{\mathrm{H}}(\overline{\mathbf{x}}),\end{aligned} \tag{3.32}$$

where $\mathbf{U}_k(\overline{\mathbf{x}}) = \mathbf{V}_k(\overline{\mathbf{x}})\mathbf{C}_k$. It is also straightforward to find $\mathbf{V}_k(\overline{\mathbf{x}})$ in Eq. (3.28) from $\mathbf{V}(\overline{\mathbf{x}})$ using the matrix inversion lemma. In Eqs. (3.29) and (3.32), it can be seen that we only need $D \times D$ matrix inversions.

The process of VI detection is presented in Algorithm 3.1, where *iter* denotes the number of iterations and ϵ denotes the accuracy requirement of detection.

Algorithm 3.1 VI Detection for Sparse Signal

Input: $\mathbf{y}$, $\mathbf{C}$, *maxiter*
Output: The distribution of $\mathbf{x}$

Initialize: ELBO
for $iter \in \{1, \ldots, maxiter\}$ **do**
  for $k \in \{1, \ldots, K\}$ **do**
    Compute $\mathbf{C}_k$, $\mathbf{v}_k(\overline{\mathbf{x}})$, $\mathbf{U}_k(\overline{\mathbf{x}})$;
    Compute $\mathbf{z}_k$, δ, $\phi_k(\overline{\mathbf{x}})$;
    Compute ℓ_k according to Eq. (3.29);
    Compute $\psi_k(x_k = 1)$ and $\psi_k(x_k = 0)$ according to Eqs. (3.30) and (3.31), respectively;
    Update $\mathbf{U}_k(\overline{\mathbf{x}})$ and $\mathbf{V}(\overline{\mathbf{x}})$ according to Eq. (3.32);
  end for
  Update ELBO(*iter*)
end for
if $\|$ ELBO($iter + 1$) – ELBO($iter$) $\| < \epsilon$ **then**
  Return result;
end if

3.3 Performance Analysis

3.3.1 Complexity Analysis

In this section, we discuss the complexity of the tree search-based approach under the assumption that the activity variables, x_k, are correctly detected. For comparison, the conventional sparse signal detection based on CS algorithm is discussed. In conventional approaches, the sparse signal detection is carried out at all groups of received signals using CS algorithm, e.g., OMP, LASSO, and AMP. The conventional approach has the total complexity as [40, 45, 46]:

$$\eta_{\text{conv}} = MO_{\text{II}}, \tag{3.33}$$

where O_{II} denotes the complexity of CS algorithm (referred as Stage 2) per group and signal.

Because the average number of inactive groups (associated with $|\mathcal{I}_k| = 0$) is given by

$$\overline{K}_{\text{ia}} = K(1-p)^D = K(1-p)^{\frac{M}{K}}, \tag{3.34}$$

the average number of signals that are removed in Stage 2 becomes $D\overline{K}_{\text{ia}} = M(1-p)^{\frac{M}{K}}$, which increases with K. However, as K increases, the complexity of Stage 1 increases. The total complexity of the TS approach as a function of D for given M can be given by

$$\begin{aligned}\eta_{\text{ts}}(D) &= KO_{\text{I}} + \left(M - M(1-p)^{\frac{M}{K}}\right) O_{\text{II}} \\ &= \frac{M}{D} O_{\text{I}} + M\left(1 - (1-p)^D\right) O_{\text{II}},\end{aligned} \tag{3.35}$$

where O_{I} denotes the complexity of activity variable detection (referred as Stage 1) per group and signal.

As a result, the complexity ratio is given by

$$\frac{\eta_{\text{ts}}(D)}{\eta_{\text{conv}}} = \left(1 - (1-p)^D\right) + \frac{O_{\text{I}}}{DO_{\text{II}}}. \tag{3.36}$$

It is indicated that the complexity ratio is influenced by the number of signals per group and the complexity ratio of Stages 1 and 2. From this, it can be seen that it is necessary to have $O_{\text{I}} < DO_{\text{II}}$ to keep the complexity ratio low, i.e., less than 1. In other words, for a computational saving using the TS approach, the complexity of Stage 1 has to be lower than the D times of that of Stage 2.

In Fig. 3.3, the complexity ratio is shown as functions of the number of signals per group, D, and the activity probability p. In general, we can see that the complexity ratio becomes low for a low activity probability p and a low value of $\frac{O_{\text{I}}}{O_{\text{II}}}$, while the complexity ratio decreases and then increases as D increases (as shown in Fig. 3.3 (a)). With $D = 4$ or 8, as a rule, the TS approach can reduce the complexity

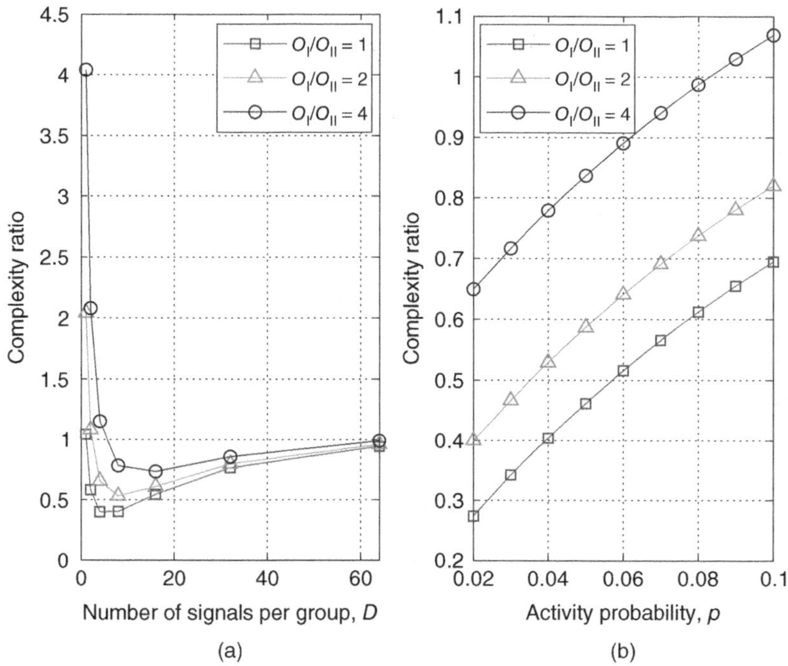

Figure 3.3 Complexity ratio: (a) as a function of D with $p = 0.04$; (b) as a function of p with $D = 8$.

(compared with the conventional one). It is observed that if $\frac{O_{\mathrm{I}}}{O_{\mathrm{II}}} = D$, there is no complexity saving using the TS approach as mentioned earlier.

3.3.2 VI Detection Performance Analysis

In this section, we mainly focus on the detection performance of Stage 1 based on the Gaussian approximation using the Neyman–Pearson test with the test statistic in Eq. (3.14). As mentioned earlier, the resulting performance can be seen as an achievable one as $\mathbf{x}_{-k}$ is required when x_k is detected. Note that the performance analysis of CAVI algorithm is not yet well known, while the analysis of the Neyman–Pearson test is to be used to see the behaviors of the performance of CAVI algorithm.

Since $\mathbf{V}_k$ is positive semidefinite, we can consider the following factorization as

$$\mathbf{V}_k = \mathbf{V}_k^{\frac{1}{2}} \mathbf{V}_k^{\frac{\mathrm{H}}{2}}. \tag{3.37}$$

Under $\mathcal{H}_0$, define

$$\mathbf{q}_k = \mathbf{V}_k^{\frac{1}{2}} \mathbf{y}. \tag{3.38}$$

Then, after some manipulations, the test statistic in Eq. (3.14) is re-written as

$$A_k = \mathbf{q}_k^{\mathrm{H}}(\mathbf{I} - \mathbf{S}_k^{-1})\mathbf{q}_k, \tag{3.39}$$

where

$$\begin{aligned}\mathbf{S}_k &= \mathbf{V}_k^{\frac{1}{2}}(\mathbf{R}_k + \mathbf{V}_k^{-1})\mathbf{V}_k^{\frac{\mathrm{H}}{2}} \\ &= \mathbf{I} + \mathbf{V}_k^{\frac{1}{2}}\mathbf{R}_k\mathbf{V}_k^{\frac{\mathrm{H}}{2}} \\ &= \mathbf{E}_k\mathbf{\Lambda}_k\mathbf{E}_k^{\mathrm{H}}.\end{aligned} \tag{3.40}$$

In Eq. (3.40), the equality is the eigenvalue decomposition of $\mathbf{S}_k$, where $\mathbf{E}_k$ is the matrix consisting of the eigenvectors and $\mathbf{\Lambda}_k$ is the diagonal matrix consisting of the eigenvalues that are denoted by $\lambda_{k,1} \geq \cdots \geq \lambda_{k,N}$.

Consequently, the characteristic function under each hypothesis is given by

$$\mathbb{E}[e^{j\omega A_k}] = \begin{cases} \prod_n \frac{1}{1-j\omega(1-1/\lambda_{k,n})}, & \text{under } \mathcal{H}_0, \\ \prod_n \frac{1}{1-j\omega(\lambda_{k,n}-1)}, & \text{under } \mathcal{H}_1. \end{cases} \tag{3.41}$$

As a result, the test statistic can also be expressed (in terms of the distribution) as

$$A_k = \sum_{n=1}^{N} \frac{\alpha_{k,n}}{2}\chi_n^2, \tag{3.42}$$

where the χ_n^2's are independent chi-squared random variables with 2 degrees of freedom and

$$\alpha_{k,n} = \begin{cases} 1 - \frac{1}{\lambda_{k,n}}, & \text{under } \mathcal{H}_0, \\ \lambda_{k,n} - 1, & \text{under } \mathcal{H}_1. \end{cases} \tag{3.43}$$

We can find the probabilities of missed detection (MD) and false alarm (FA) in Stage 1 for a given threshold τ using the distributions from the characteristic functions in Eq. (3.41).

For an alternative performance measure that is independent of τ, the J-divergence [42] can be used, which is a measure of the distance between the distributions associated with $\mathcal{H}_0$ and $\mathcal{H}_1$. From Eq. (3.41), the J-divergence is given by

$$\begin{aligned}J_k &= \mathbb{E}[A_k \mid \mathcal{H}_1] - \mathbb{E}[A_k \mid \mathcal{H}_0] \\ &= \mathrm{Tr}(\mathbf{E}_k(\mathbf{I} - \mathbf{\Lambda}_k^{-1})\mathbf{E}_k^{\mathrm{H}}\mathbf{E}_k\mathbf{\Lambda}_k\mathbf{E}_k^{\mathrm{H}}) - \mathrm{Tr}(\mathbf{E}_k(\mathbf{I} - \mathbf{\Lambda}_k^{-1})\mathbf{E}_k^{\mathrm{H}}\mathbf{I}) \\ &= \mathrm{Tr}(\mathbf{\Lambda}_k + \mathbf{\Lambda}_k^{-1} - 2\mathbf{I}) \\ &= \sum_{n=1}^{N}\left(\lambda_{k,n} + \frac{1}{\lambda_{k,n}} - 2\right) \\ &\approx \sum_{n=1}^{N}\lambda_{k,n} - N.\end{aligned} \tag{3.44}$$

In general, as the J-divergence increases, a better detection performance is expected. Since $\sum_{n=1}^{N} \lambda_{k,n} = \text{Tr}(\mathbf{S}_k)$, from Eq. (3.40), it can be obtained that

$$\sum_{n=1}^{N} \lambda_{k,n} - N = \text{Tr}\left(\mathbf{V}_k^{\frac{1}{2}} \mathbf{R}_k \mathbf{V}_k^{\frac{H}{2}}\right)$$
$$= \text{Tr}(\mathbf{V}_k \mathbf{R}_k)$$
$$= \text{Tr}\left(\left(\sum_{q \neq k} \mathbf{R}_q x_q + N_0 \mathbf{I}\right)^{-1} \mathbf{R}_k\right). \tag{3.45}$$

Lemma 3.1 *Suppose that* $\mathbf{x}_{-k} \geq \mathbf{x}'_{-k}$, *i.e.,* $x_q \geq x'_q$ *for* $q \in \{1, \dots, K\} \backslash k$. *In addition, denote by* $\mathbf{V}_k(\mathbf{x}_{-k})$ *for* $\mathbf{V}_k$ *that is associated with* $\mathbf{x}_{-k}$. *Then, we have*

$$\text{Tr}(\mathbf{V}_k(\mathbf{x}_{-k})\mathbf{R}_k) \leq \text{Tr}(\mathbf{V}_k(\mathbf{x}'_{-k})\mathbf{R}_k). \tag{3.46}$$

Proof: Since $\mathbf{x}_{-k} \geq \mathbf{x}'_{-k}$, it can be shown that

$$\mathbf{V}_k(\mathbf{x}_{-k}) = \left(\sum_{q \neq k} \mathbf{R}_q x_q + N_0 \mathbf{I}\right)^{-1}$$
$$= \left(\sum_{q \neq k} \mathbf{R}_q (x'_q + x_q - x'_q) + N_0 \mathbf{I}\right)^{-1}$$
$$= (\mathbf{V}_k^{-1}(\mathbf{x}'_{-k}) + \mathbf{E})^{-1}, \tag{3.47}$$

where $\mathbf{E} = \sum_{q \neq k} \mathbf{R}_q (x_q - x'_q)$, which is a positive semidefinite matrix. Thus, using the matrix inversion lemma, it can be shown that

$$\mathbf{V}_k(\mathbf{x}_{-k}) - \mathbf{V}_k(\mathbf{x}'_{-k}) = (\mathbf{V}_k^{-1}(\mathbf{x}'_{-k}) + \mathbf{E})^{-1} - \mathbf{V}_k(\mathbf{x}'_{-k})$$
$$= \mathbf{V}_k(\mathbf{x}'_{-k})(-\mathbf{I} - \mathbf{E}\mathbf{V}_k(\mathbf{x}'_{-k}))^{-1} - \mathbf{V}_k(\mathbf{x}'_{-k})$$
$$\preccurlyeq \mathbf{0}. \tag{3.48}$$

From this, since $\mathbf{V}_k(\mathbf{x}'_{-k}) - \mathbf{V}_k(\mathbf{x}_{-k})$ is positive semidefinite, it can be shown that

$$\text{Tr}(\mathbf{V}_k(\mathbf{x}'_{-k})\mathbf{R}_k) - \text{Tr}(\mathbf{V}_k(\mathbf{x}_{-k})\mathbf{R}_k)$$
$$= \text{Tr}\left((\mathbf{V}_k(\mathbf{x}'_{-k}) - \mathbf{V}_k(\mathbf{x}_{-k}))\mathbf{R}_k\right) \geq 0, \tag{3.49}$$

which completes the proof.

From Lemma 3.1, we can easily see that the performance can be improved if a small number of groups are active and the J-divergence can be bounded as

$$J_k \approx \sum_{n=1}^{N} \lambda_{k,n} - N \geq \text{Tr}\left(\left(\sum_{q \neq k} \mathbf{R}_q + N_0 \mathbf{I}\right)^{-1} \mathbf{R}_k\right). \tag{3.50}$$

As a special case, suppose that there is no inter-group interference or each $\mathbf{C}_k$ is orthogonal to the others. Then, we have

$$\text{Range}(\mathbf{C}_k) \perp \text{Range}(\mathbf{C}_{k'}),\ k \neq k'. \tag{3.51}$$

From Eq. (3.51), we have

$$\mathbf{S}_k = \mathbf{I} + \frac{1}{N_0}\mathbf{R}_k = \mathbf{I} + \frac{\beta}{N_0}\mathbf{C}_k\mathbf{C}_k^{\text{H}}, \tag{3.52}$$

and

$$\text{Tr}(\mathbf{C}_k\mathbf{C}_k^{\text{H}}) = \sum_{m\in\mathcal{I}_k} ||\mathbf{c}_m||^2 = |\mathcal{I}_k| = D, \tag{3.53}$$

if $||\mathbf{c}_m||^2 = 1$ for all m. As a result, it can be shown that

$$\begin{aligned} J_k &\approx \text{Tr}(\mathbf{S}_k) - N \\ &= \frac{\beta}{N_0}\ \text{Tr}(\mathbf{C}_k\mathbf{C}_k^{\text{H}}) = \frac{\beta D}{N_0}. \end{aligned} \tag{3.54}$$

This indicates that as D increases (or K decreases) for a fixed M, a better performance can be obtained. Note that in order to satisfy Eq. (3.51), the rank of $\mathbf{C}_k$ might be

$$\overline{D} = \frac{N}{K} < D\left(= \frac{M}{K}\right), \tag{3.55}$$

which means that the $\mathbf{c}_m$'s in $\mathcal{I}_k$ can be highly correlated (if $\overline{D} \ll D$), which can lead to a poor performance of Stage 2.

Remark 3.1 In general, without any additional properties of $\mathbf{C}$ or $\{\mathbf{R}_k\}$, it is difficult to find the J-divergence or the trace of $\mathbf{S}_k$. However, when M and N are sufficiently large, we can consider a large-system analysis to find the trace of $\mathbf{S}_k$ as follows.

Lemma 3.2 *Denote by I the number of non-zero elements of $\mathbf{x}_{-k}$, i.e., $I = ||\mathbf{x}_{-k}||_1$. In addition, let $\kappa_I = \frac{DI}{N}$. Assume that each element of $\mathbf{C}$ is an independent random variable with zero-mean and variance $\frac{1}{N}$. For a given I, suppose that N increases with a fixed ratio κ_I. Then, we have*

$$\sum_{n=1}^{N} \lambda_{k,n} - N \to D\gamma(\kappa_I),\ N \to \infty, \tag{3.56}$$

where

$$\gamma(\kappa) = \frac{(1-\kappa)\beta}{2N_0} - \frac{1}{2} + \sqrt{\left(\frac{(1-\kappa)\beta}{2N_0}\right)^2 + \frac{(1+\kappa)\beta}{2N_0} + \frac{1}{4}}. \tag{3.57}$$

Proof: From Eq. (3.45), it can be shown that

$$\begin{aligned}
\sum_{n=1}^{N} \lambda_{k,n} - N &= \mathrm{Tr}\left(\left(\sum_{q \neq k} \mathbf{R}_q x_q + N_0 \mathbf{I} \right)^{-1} \mathbf{R}_k \right) \\
&= \mathrm{Tr}\left(\beta \mathbf{C}_k \left(\sum_{q \neq k} \mathbf{R}_q x_q + N_0 \mathbf{I} \right)^{-1} \mathbf{C}_k^{\mathrm{H}} \right) \\
&= \beta \sum_{m \in \mathcal{I}_k} \mathbf{c}_m^{\mathrm{H}} \left(\sum_{q \neq k} \mathbf{R}_q x_q + N_0 \mathbf{I} \right)^{-1} \mathbf{c}_m \\
&= \sum_{m \in \mathcal{I}_k} \mathbf{c}_m^{\mathrm{H}} \left(\sum_{q \neq k} \mathbf{C}_q \mathbf{C}_q^{\mathrm{H}} x_q + \frac{N_0}{\beta} \mathbf{I} \right)^{-1} \mathbf{c}_m . \qquad (3.58)
\end{aligned}$$

Each term in Eq. (3.58) can be seen as the signal-to-interference-plus-noise ratio (SINR) of MIMO systems, where it is shown that the asymptotic SINR (for $N \to \infty$) depends on $\frac{\beta}{N_0}$ and κ_I. In particular, the asymptotic SINR becomes Eq. (3.57) with $\kappa = \kappa_I$. In addition, since each $\mathbf{c}_m$ is independent of the others, we have $D \times$ asymptotic SINR, which is given in Eq. (3.56). This completes the proof.

Theorem 3.1 *Suppose that the x_k's are i.i.d. with $\Pr(x_k = 1) = 1 - (1-p)^D$ and $\Pr(x_k = 0) = (1-p)^D$. For a large system, where $N \to \infty$ with a fixed ratio $\frac{N}{D}$, we can show that*

$$\begin{aligned}
J_k &\approx \mathbb{E}\left[\sum_{n=1}^{N} \lambda_{k,n} \right] - N \\
&\to \mathbb{E}_I[D\gamma(\kappa_I)] \\
&= D \sum_{i=0}^{K-1} \gamma(\kappa_i) \mathbb{P}_{K-1}(i), \qquad (3.59)
\end{aligned}$$

where $\mathbb{P}_{K-1}(i) = \binom{K-1}{i} \left(1 - (1-p)^D\right)^i (1-p)^{D(K-1-i)}$.

Proof: Since $\frac{N}{D}$ is fixed, $\kappa_I = \frac{DI}{N}$ is also fixed when $N \to \infty$. Here, I is a binomial random variable with parameters $K-1$ and $1-(1-p)^D$, because the x_k's are independent binary random variables. In particular, $\Pr(I = i) = \mathbb{P}_{K-1}(i)$. Thus, we can show that

$$\begin{aligned}
\mathbb{E}\left[\sum_{n=1}^{N} \lambda_{k,n} \right] &= \mathbb{E}\left[\mathbb{E}\left[\sum_{n=1}^{N} \lambda_{k,n} \,|\, I \right] \right] \\
&= \sum_{i=0}^{K-1} \mathbb{E}\left[\sum_{n=1}^{N} \lambda_{k,n} \,|\, I \right] \mathbb{P}_{K-1}(i). \qquad (3.60)
\end{aligned}$$

Then, substituting Eq. (3.56) into Eq. (3.60), Eq. (3.59) can be obtained, which completes the proof.

After some manipulations, we can see that $\gamma(\kappa)$ is convex in κ (and decreases with κ). Thus, thanks to Jensen's inequality, we can also have

$$\mathbb{E}_I[D\gamma(\kappa_I)] \geq D\gamma(\overline{\kappa}), \tag{3.61}$$

where

$$\begin{aligned}\overline{\kappa} &= \mathbb{E}_I[\kappa_I] \\ &= \frac{D}{N}(K-1)(1-(1-p)^D) \\ &\approx \frac{(M-D)(1-e^{-pD})}{N}.\end{aligned} \tag{3.62}$$

From the first and second derivatives, which are given by

$$\frac{d\,(M-D)(1-e^{-pD})}{dD} = e^{-pD}((M-D)p+1)-1, \tag{3.63}$$

and

$$\frac{d^2\,(M-D)(1-e^{-pD})}{dD^2} = -pe^{-pD}((M-D)p+2), \tag{3.64}$$

we can see that $\overline{\kappa}$ is concave in D and has a $\cap$-shape. Consequently, there exists a $D \in [1, M]$ maximizing κ, which leads to the minimum of $\gamma(\overline{\kappa})$, i.e., the worst performance. In other words, in order to keep a good performance (or a large J-divergence), D has to be close to M or 1. In addition, according to Eq. (3.62), the increase of p leads to the increase of $\overline{\kappa}$, as there are more active signals. From this, we can clearly see that the approximation of J-divergence decreases with p.

In Fig. 3.4, the J-divergence and its approximation in Eq. (3.59) are shown as functions of D (with $p = 0.04$) and p (with $D = 8$), when $N = 128$, $M = 512$, and $\frac{E_s}{N_0} = 20$ dB. Note that as mentioned earlier, $\sum_n \lambda_{k,n} - N$ is actually an analysis of the J-divergence, while the behavior of the J-divergence in terms of D can be seen using the approximation that has a closed-form expression as in Eq. (3.59). Since we use the scaling transform and the large-system analysis to approximate the J-divergence, the simulations and analysis performance have a gap. The J-divergence and its approximation become L-shape functions of D as shown in Fig. 3.4 (a), which indicates that the increase of D in general degrades the performance. In addition, the performance becomes worse as p increases (that increases the interference level) as shown in Fig. 3.4 (b). Consequently, we can see that a large D is not desirable in terms of the performance as well as the complexity, while a small, but not too small D (e.g., $D = 4$ or 8), might be a good choice.

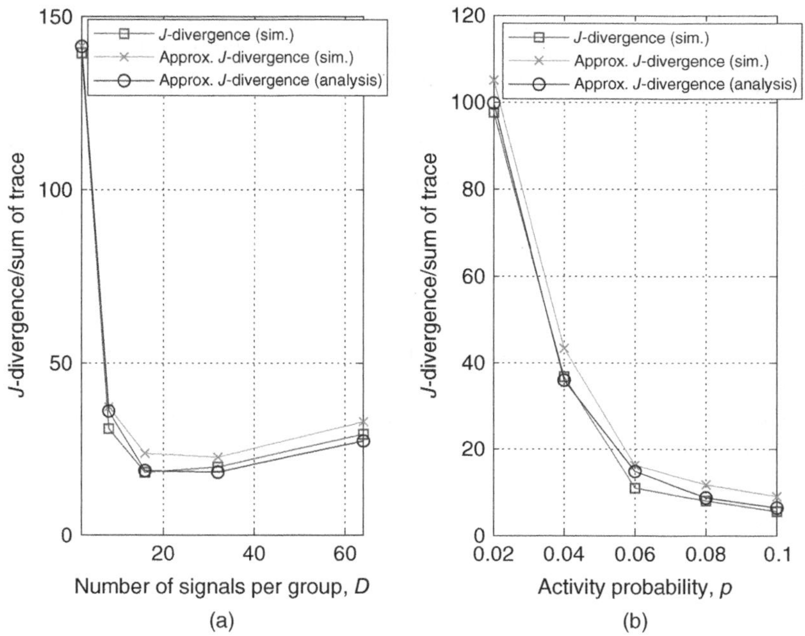

Figure 3.4 J-divergence and its approximation in Eq. (3.59) when $N = 128$, $M = 512$, and $\frac{E_s}{N_0} = 20$ dB: (a) as a function of D with $p = 0.04$; (b) as a function of p with $D = 8$.

3.4 Simulation Results

In this section, we present simulation results. It is assumed that each element of $\mathbf{C}$ is independent zero-mean circular symmetric complex Gaussian (CSCG) with variance $\frac{1}{N}$. In addition, we assume that b_m is a quadrature phase shift keying (QPSK) symbol, i.e., $b_m \in \{\pm 1, \pm j\}$, where $j = \sqrt{-1}$. The signal-to-noise ratio (SNR) is defined as $\text{SNR} = \frac{E_s}{N_0}$, where the symbol energy is $E_s = 2$ for QPSK.

Firstly, we mainly present simulation results for the performance of CAVI algorithm for Stage 1. We consider the decision rule as

$$\psi_k^*(x_k = 1) \underset{\mathcal{H}_0}{\overset{\mathcal{H}_1}{\gtrless}} p_\tau, \tag{3.65}$$

where $p_\tau \in (0, 1)$ is a decision threshold. In addition, we consider two different values for p_τ, which are 0.5 and $\Pr(x_k = 1) = 1 - (1 - p)^D$. If $p_\tau = 0.5$, we decide that $x_k = 1$ when $\psi_k^*(x_k = 1) > \psi_k^*(x_k = 0)$. On the other hand, if $p_\tau = \Pr(x_k = 1) = 1 - (1 - p)^D$, we take into account the a priori probability to make a decision, where

$x_k = 1$ is decided when $\psi_k^*(x_k = 1) \geq \Pr(x_k = 1) = 1 - (1-p)^D$. For simulations, the number of iterations for the CAVI algorithm is set to 4. Probabilities of FA and MD (for given p_τ) for Stage 1 are obtained by simulations. Here, an event of FA or MD is obtained for each group.

Figure 3.5 shows the performance of Stage 1 for various SNRs with $M = 512$, $(K, D) = (128, 4)$, $N = 128$, $p = 0.02$, and $p_\tau \in (\Pr(x_k = 1), 0.5)$. In general, we can see that the probabilities of FA and MD of the CAVI algorithm decrease with the SNR in Fig. 3.5 (a). This indicates that a better performance can be achieved with a higher SNR, which is also confirmed by the J-divergence that increases with the SNR as shown in Fig. 3.5 (b).

Figure 3.6 shows the performance of Stage 1 with different activity probabilities p, when $M = 512$, $(K, D) = (128, 4)$, $N = 128$, SNR= 16 dB, and $p_\tau \in (\Pr(x_k = 1), 0.5)$. It shows that probabilities of FA and MD of the CAVI algorithm increase with the activity probability as the larger activity probability results in a higher interference level, which can also be seen in Fig. 3.6 (b) that the J-divergence decreases with p increases.

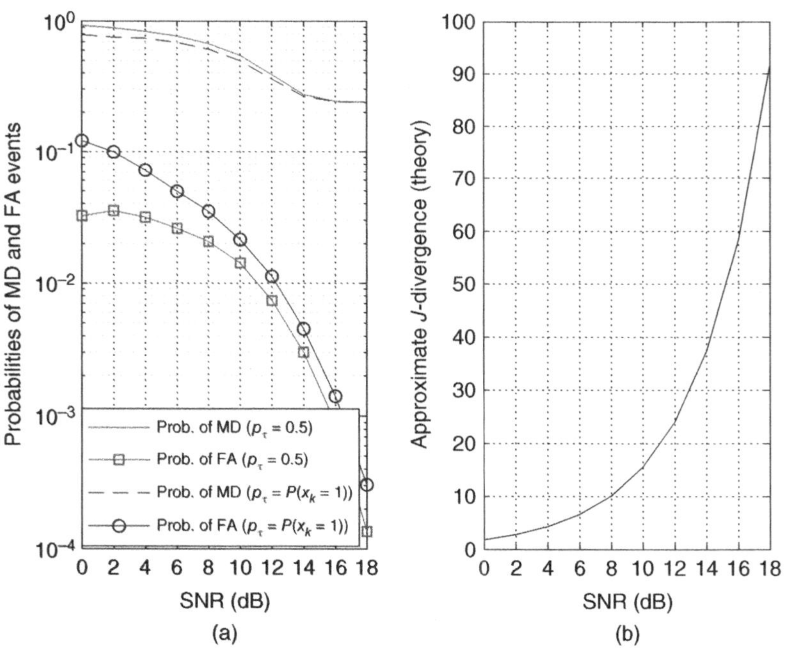

Figure 3.5 Performance of Stage 1 for various SNRs with $M = 512$, $(K, D) = (128, 4)$, $N = 128$, $p = 0.02$, and $p_\tau \in (\Pr(x_k = 1), 0.5)$: (a) the probabilities of FA and MD of the CAVI algorithm as functions of SNR; (b) J-divergence as a function of SNR.

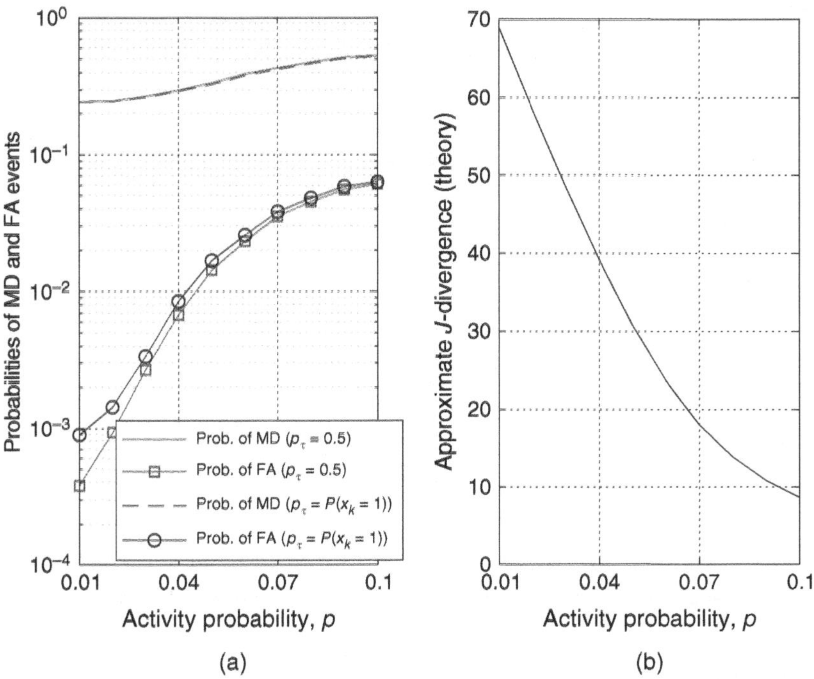

Figure 3.6 Performance of Stage 1 for various values of p with $M = 512$, $(K, D) = (128, 4)$, $N = 128$, SNR $= 16$ dB, and $p_\tau \in (\Pr(x_k = 1), 0.5)$: (a) the probabilities of FA and MD of the CAVI algorithm as functions of p; (b) J-divergence as a function of p.

The impact of D on the performance is shown in Fig. 3.7, where the performance is shown for various values of D with $M = 512$, $N = 128$, $p = 0.02$, SNR $= 16$ dB, and $p_\tau \in (\Pr(x_k = 1), 0.5)$. As shown in Fig. 3.7 (a), we see that the MD and FA probabilities of the CAVI algorithm first increase and then tend to be stable as D increases. Similarity, the J-divergence in Fig. 3.7 (b) has a L-shape, which indicates that the performance becomes worse and saturated as D increases. Consequently, according to Fig. 3.7 as well as Fig. 3.3, it is necessary to keep D small, but not too small for a low complexity. From them, in most cases, it seems that $D = 4$ is a reasonable choice.

Figure 3.8 shows the impact of N on the performance of Stage 1 with $M = 512$, $(K, D) = (128, 4)$, $p = 0.02$, SNR $= 16$ dB, and $p_\tau \in (\Pr(x_k = 1), 0.5)$. As we can see that a better performance can be achieved with an increasing N. As shown in Fig. 3.8 (b), the J-divergence increases as N increases, which indicates that the performance becomes better.

As discussed above, we decide active and inactive groups with the activity variables in Stage 1. For Stage 2, the signals associated with inactive groups are

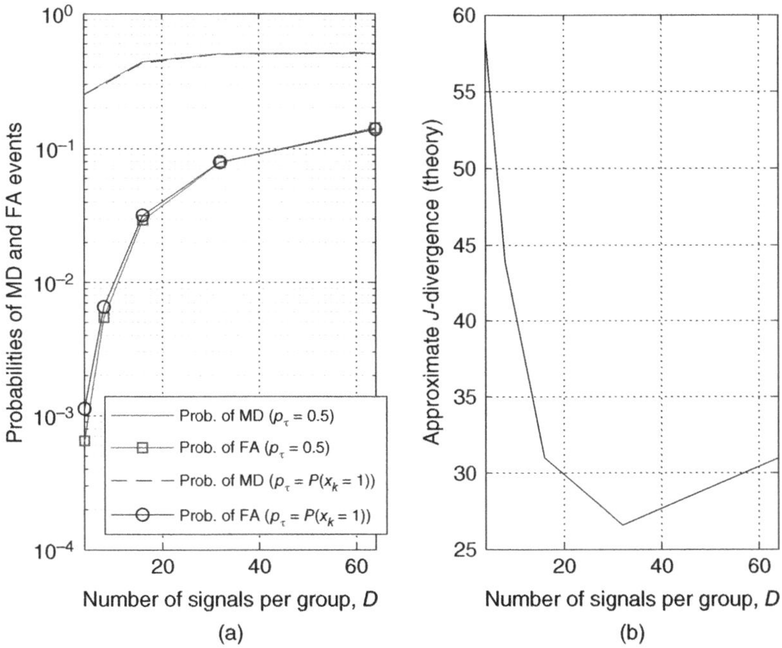

Figure 3.7 Performance of Stage 1 for various values of D with $M = 512$, $N = 128$, $p = 0.02$, SNR = 16 dB, and $p_\tau \in (\Pr(x_k = 1), 0.5)$: (a) the probabilities of FA and MD of the CAVI algorithm as functions of D; (b) J-divergence as a function of D.

removed and the sparse signal detection is carried out with the signals belonging to active groups. In this way, the complexity at the receiver becomes lower. In order to show the performance of the TS approach, we consider the OMP algorithm, the LASSO algorithm, and the AMP algorithm to detect the sparse signal, respectively. For convenience, the TS approach with OMP, LASSO, or AMP for Stage 2 is referred to as TS-OMP, TS-LASSO, or TS-AMP, respectively. In addition, for performance comparisons, we consider a SS approach with OMP, LASSO, or AMP, which is referred to as SS-OMP, SS-LASSO, or SS-AMP, respectively.

Figure 3.9 shows the performance of the TS approach and the SS approach for various SNRs with $M = 512, (K, D) = (128, 4), N = 128, p = 0.02$, and $p_\tau = 0.5$. It is shown that bit error rate (BER) decreases with SNR increases, which indicates that a high SNR is desirable for a good detection performance. Since these approaches have different diversity, the performance of detection may be different. However, the detection performance of these approaches becomes indistinguishable at some points as it is influenced by the noise. In general, the TS-LASSO and the TS-AMP algorithms can achieve better detection performance than the TS-OMP algorithm.

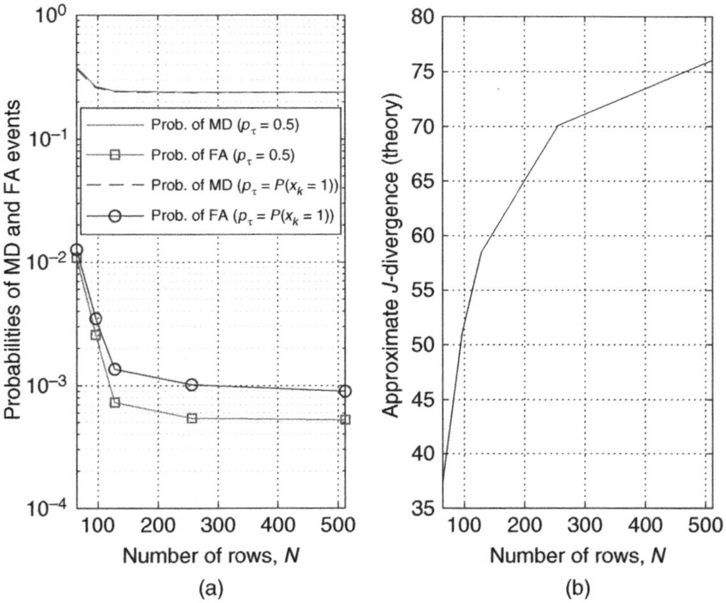

Figure 3.8 Performance of Stage 1 for various values of N with $M = 512$, $(K, D) = (128, 4)$, $p = 0.02$, SNR = 16 dB, and $p_\tau \in (\Pr(x_k = 1), 0.5)$: (a) the probabilities of FA and MD of the CAVI algorithm as functions of N; (b) J-divergence as a function of N.

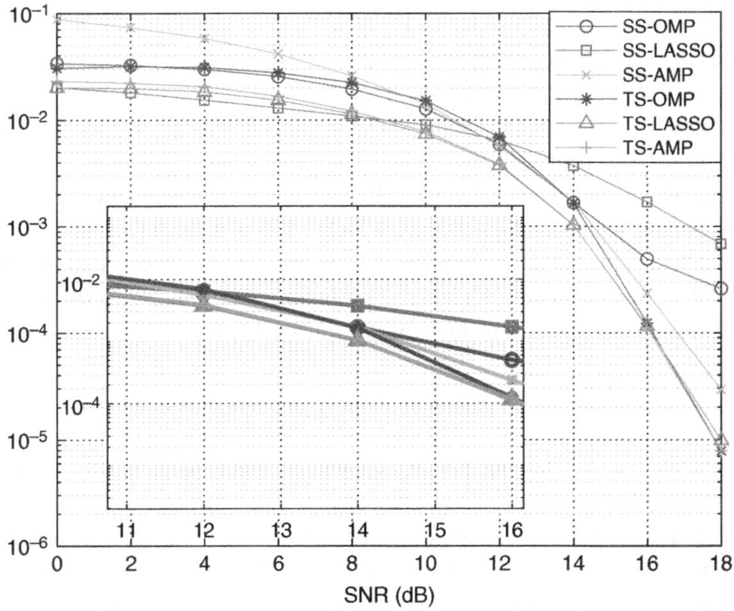

Figure 3.9 Performance of TS approach and SS approach for various values of SNR with $M = 512$, $(K, D) = (128, 4)$, $N = 128$, $p = 0.02$, and $p_\tau = 0.5$.

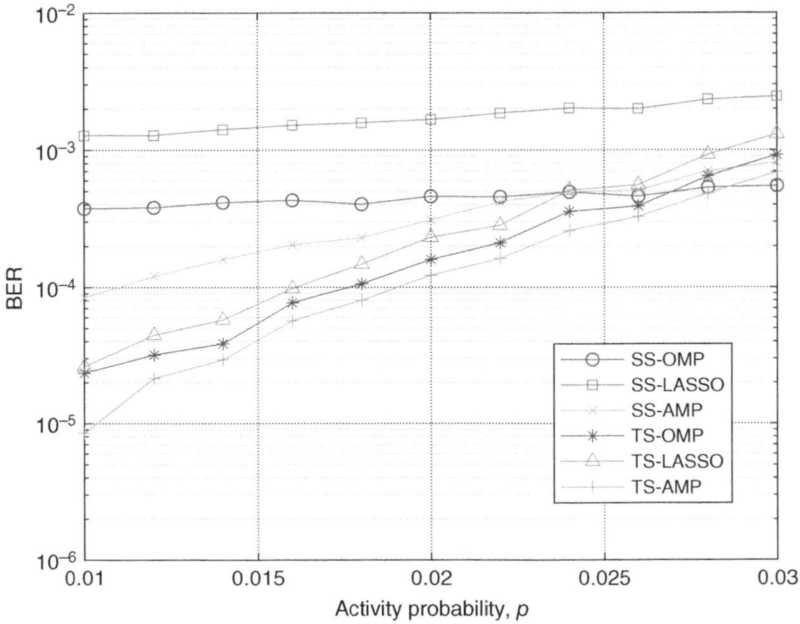

Figure 3.10 Performance of TS approach and SS approach for various values of p with $M = 512$, $(K, D) = (128, 4)$, $N = 128$, SNR = 16 dB, and $p_\tau = 0.5$.

Besides, we can see that the TS approach significantly improves the detection performance when SNR becomes higher.

Figure 3.10 shows the performance of the TS approach and the SS approach with different activity probabilities p, when $M = 512$, $(K, D) = (128, 4)$, $N = 128$, SNR = 16 dB, and $p_\tau = 0.5$. In general, the BER of both approaches increases with the activity probability as the larger activity probability results in the higher interference level. In addition, the performance of TS approach has improvement compared with SS approach, when the activity probability is small. Since the activity probability of sparse signal can be low, the TS approach can achieve a good signal detection performance.

Figure 3.11 shows the performance of the TS approach and the SS approach for various values of N with $M = 512$, $(K, D) = (128, 4)$, SNR = 16 dB, $p = 0.02$, and $p_\tau = 0.5$. We can see that BER decreases with the number of outputs, which indicates that the larger N leads to a better detection performance. It also shows that the BER of three algorithms first declines rapidly and then tends to be stable. Compared two approaches, we can see that the algorithms of TS approach achieve lower BER than those of SS approach especially at a large number of outputs.

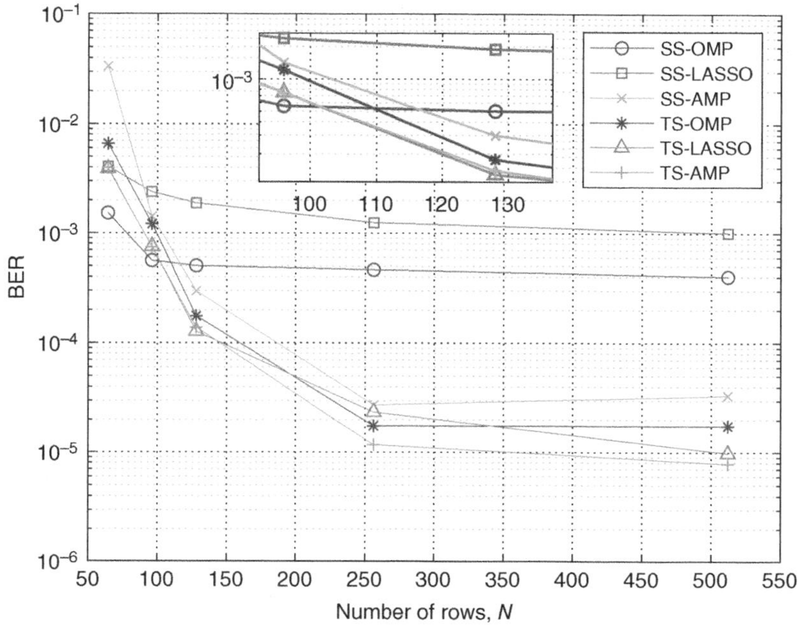

Figure 3.11 Performance of TS and SS approach for various values of N with $M = 512$, $(K, D) = (128, 4)$, SNR = 16 dB, $p = 0.02$, and $p_\tau = 0.5$.

3.5 Conclusion and Remarks

In this chapter, we introduced a tree search-based approach to detect sparse signals for NGMA schemes, where the received signals were divided into groups. Then, the activity variables of aggregated signals were decided and the sparse signal detection could be carried out at the signals belonging to active groups, where the VI was applied to the activity variable detection. In addition, the J-divergence was introduced to measure the distance between the distributions of activity variable detection, where the approximate expression of J-divergence was given.

To further improve the performance of IoT systems, different collided signal detection and estimation methods are designed and then adopted, which will be introduced in Chapters 4 and 5.

4

Collided Signal Detection for Multiple Access

- To communicate with a base stations (BS), an Internet of Things (IoT) device must initiate an random access (RA) procedure via sending a preamble that is randomly selected from a predefined set through a shared physical random access channel (PRACH) [47, 48]. For massive access, the case that multiple IoT devices choose the same preamble namely preamble collision can occur frequently, while the data packet from the device experiencing collision may not be successfully decoded at the BS. Therefore, overcoming congestion and overload of PRACH is highly desirable, especially for massive connectivity to support machine-type communication (MTC) communications. In this chapter, we introduce collided signal detection approaches for massive access-based grant-free RA. Based on the channel hardening and favorable propagation characteristics of massive multiple-input multiple-output (MIMO), collided signals processed at the BS can be viewed as a variation of superposition modulation, and are to be recovered by successive interference cancellation (SIC) techniques [49].

4.1 System Model

We consider a single-cell massive MIMO network for massive access depicted in Fig. 4.1. Suppose that a BS is equipped with M antennas and there are a large number of IoT devices that are equipped with a single antenna for uplink transmissions, called user equipments (UEs). For a typical massive access scenario, the total number of UEs in a cell can be numerous. However, the number of UEs that attempt to access the BS may not be large in a certain time interval [50], due to the sparse user activity in most IoT and MTC scenarios. A UE with a data packet to transmit is called an active UE. Let K denote the number of active UEs attempting to initiate an RA procedure over the same channel.

IoT Signal Detection, First Edition. Rui Han, Jingjing Wang, Lin Bai, and Jianwei Liu.

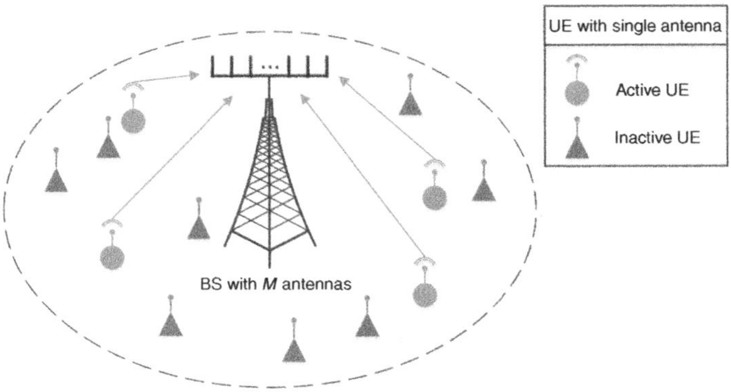

Figure 4.1 A typical uplink massive MIMO System.

In addition, $\mathbf{h}_k = \sqrt{\ell_k}\mathbf{v}_k \in \mathbb{C}^{M\times 1}$ denotes the channel vector from the kth active UE to the BS, where ℓ_k is the large-scale path loss and $\mathbf{v}_k \sim \mathcal{CN}\left(\mathbf{0}, \mathbf{I}_M\right)$ denotes the uncorrelated Rayleigh fading.

Denote by $l(k)$ the index of the preamble chosen by the kth active UE. Then, the received signal in the space-time domain at the BS is given by

$$\mathbf{X} = \sum_{k=1}^{K} \sqrt{P_k}\mathbf{h}_k\mathbf{c}_{l(k)} + \mathbf{N}_1 \in \mathbb{C}^{M\times N}. \tag{4.1}$$

Here, $\mathbf{N}_1$ represents the background noise where each element is an independent circularly symmetric complex Gaussian (CSCG) random variable with zero-mean and variance N_0, i.e., $\left[\mathbf{N}_1\right]_{m,n} \sim \mathcal{CN}(0, N_0)$, and P_k is the transmission power for the kth UE.

Denote by $\mathbf{d}_k$ the data packet transmitted by the kth UE. We assume that all the active UEs have data packet of the same length D. Then, the received signal at the BS is given by

$$\mathbf{Y} = \sum_{k=1}^{K} \sqrt{P_k}\mathbf{h}_k\mathbf{d}_k^{\mathrm{T}} + \mathbf{N}_2 \in \mathbb{C}^{M\times D}, \tag{4.2}$$

where $\mathbf{N}_2$ is the background noise with $\left[\mathbf{N}_2\right]_{m,n} \sim \mathcal{CN}(0, N_0)$.

Specifically, the large scale path loss coefficient ℓ_k is modeled as

$$\ell_k = C\left(R_k\right)^{-\alpha}, \tag{4.3}$$

where C is a constant with respect to reference distance and carrier frequency, R_k is the distance between the kth UE and the BS, and α is the path loss exponent.

Next, we model the location of the active UE as an independent homogeneous Poisson point process and the distance R_k follows, a Rayleigh distribution with mean $0.5\sqrt{1/\lambda}$, where λ is the density of BSs measured in km^2 [51, 52]. Therefore, the probability density function of R_k (in kilometer) is given by

$$f_r(r) = 2\pi\lambda r e^{-\pi\lambda r^2}, \tag{4.4}$$

and the probability density function of R_k (in meter) is $\overline{f}_r(r) = \frac{1}{1000} f_r\left(\frac{r}{1000}\right)$.

The transmission power P_k is further computed as

$$P_k = P_t(\ell_k)^{-\epsilon}, \tag{4.5}$$

where P_t denotes the open loop transmission power and $\epsilon \in [0, 1]$ denotes the path loss compensation fraction. Fractional power control (FPC) is used in the long-term evolution (LTE) systems in uplink transmission to compensate for path loss, while it can also be employed in the scenario where no coordination between UEs and the BS is needed, since UEs can derive their own transmission power from the system information broadcast by the BS [53].

Generally, attenuation of transmission power with distance is beneficial to keep the colliding UEs from obtaining the same channel gains, while the collided signals can hardly obtain similar receive power. If the superimposed symbols are weighted by similar values formed by their receive power, the detection is ambiguous, and correct recovery may not be available [54]. Hence, full power control, i.e., $\epsilon = 1$, should be avoided such that the receive power from different UEs at the BS can be distinguishable.

For RA, we assume that there is a pool of orthogonal preambles denoted by $C = \{\mathbf{c}_1, \dots, \mathbf{c}_L\} \in \mathbb{C}^{1\times N}$, where L and N represent the number and the length of preambles, respectively. Therefore, we have $\mathbf{c}_i\mathbf{c}_j^{\mathrm{H}} = 0$, for $i \neq j$. Besides, near-orthogonal preambles can be generated by Zadoff–Chu sequences such that $|\mathbf{c}_i|^2 = N$, $i \in \{1, \dots, L\}$. At the BS, the preamble is used to estimate the channel vector $\mathbf{h}_k$, that will be used for coherent detection on the following data packet [55]. For example, to estimate the channel vector of the active UE that chooses the lth preamble, we consider the output of the correlator with $\mathbf{c}_l$ as

$$\mathbf{g}_l = \frac{\mathbf{X}\mathbf{c}_l^{\mathrm{H}}}{\left\|\mathbf{c}_l^{\mathrm{H}}\right\|^2} = \frac{1}{N}\mathbf{X}\mathbf{c}_l^{\mathrm{H}}. \tag{4.6}$$

Since the preambles are assumed to be orthogonal, we have

$$\mathbf{g}_l = \sum_{k\in\zeta_l} \sqrt{P_k}\mathbf{h}_k + \tfrac{1}{N}\mathbf{N}_1\mathbf{c}_l^{\mathrm{H}}, \tag{4.7}$$

where ζ_l denotes the index set of the active UEs that choose the lth preamble.

Obviously, if $\mathbf{g}_l$ is used for channel estimation, we have three different cases as follows:

- **Case 1:** $|\zeta_l| = 0$. In this case, $\mathbf{g}_l$ is a noise vector given by $\mathbf{g}_l = \frac{1}{N}\mathbf{N}_1\mathbf{c}_l^{\mathrm{H}}$ and the BS detects no active UEs that choose $\mathbf{c}_l$.

- **Case 2:** $|\zeta_l| = 1$. In this case, the preamble $\mathbf{c}_l$ is uniquely chosen by the kth UE and $\mathbf{g}_l$ is given by $\mathbf{g}_l = \sqrt{P_k}\mathbf{h}_k + \frac{1}{N}\mathbf{N}_1\mathbf{c}_l^{\mathrm{H}}$, where $\{k\} = \{\zeta_l\}$. Then, the channel vector $\mathbf{h}_k$ can be accurately estimated using $\mathbf{g}_l$.
- **Case 3:** $|\zeta_l| \geq 2$. In this case, the preamble $\mathbf{c}_l$ is chosen by two or more UEs and $\mathbf{g}_l$ is given by $\mathbf{g}_l = \sum_{k\in\zeta_l} \sqrt{P_k}\mathbf{h}_k + \frac{1}{N}\mathbf{N}_1\mathbf{c}_l^{\mathrm{H}}$. Note that $\mathbf{g}_l$ includes a superposition of channel vectors, therefore, the channel vectors cannot be correctly estimated using $\mathbf{g}_l$. The corresponding channel estimate generated from $\mathbf{g}_l$ thus becomes a superposition of multiple channel vectors of the active UEs that choose $\mathbf{c}_l$ (including the noise).

Besides, with the beamforming vector $\boldsymbol{\varphi}_l$ generated from $\mathbf{g}_l$, the BS detects data packets with the signal given by

$$\mathbf{y}_l = \boldsymbol{\varphi}_l\mathbf{Y}. \tag{4.8}$$

Specifically, the beamforming vector of conjugate beamforming (CB) is $\boldsymbol{\varphi}_l = \mathbf{g}_l^{\mathrm{H}}$ and the beamforming vector of zero-forcing beamforming (ZFB) is the lth row of the matrix $\left(\mathbf{A}^{\mathrm{H}}\mathbf{A}\right)^{-1}\mathbf{A}^{\mathrm{H}}$, where $\mathbf{A}$ is the channel estimation matrix defined as $\mathbf{A} = \left[\mathbf{g}_1, \mathbf{g}_2, \dots, \mathbf{g}_{S+1}\right] \in \mathbb{C}^{M\times(S+1)}$ [56, 57]. Here, $S+1$ denotes the number of preambles totally picked by the K co-channel UEs.

4.2 Automatic Modulation Classification-based Detection

Under the circumstance that the lth preamble $\mathbf{c}_l$ is employed with the received signal $\mathbf{X}$ as an input, Cases 1–3 can be classified as depicted in Fig. 4.2, where the procedure is introduced in detail as follows.

4.2.1 Preamble Sequence Detection

Upon a UE performing a RA procedure to access the network, the BS detects the preambles in use by calculating the cross-correlation between the received signal $\mathbf{X}$ and L available preambles, after which the corresponding channels are to be observed by the BS. Here, we consider an energy detector to decide Case 1, while preamble $\mathbf{c}_l$ is detected if $\left|\mathbf{X}\mathbf{c}_l^{\mathrm{H}}\right|^2$ is larger than a predefined threshold.

The decision of Case 1 can be formulated into a binary hypothesis testing problem. $\mathcal{H}_0$ and $\mathcal{H}_1$ refer to the hypotheses of signal absence and presence, respectively, which are given by

$$\begin{aligned}
\mathcal{H}_0 &: \mathbf{x} = \mathbf{N}_1\mathbf{c}_l^{\mathrm{H}}, \\
\mathcal{H}_1 &: \mathbf{x} = \sqrt{P_k}\mathbf{h}_k|\mathbf{c}_l|^2 + \mathbf{N}_1\mathbf{c}_l^{\mathrm{H}}.
\end{aligned} \tag{4.9}$$

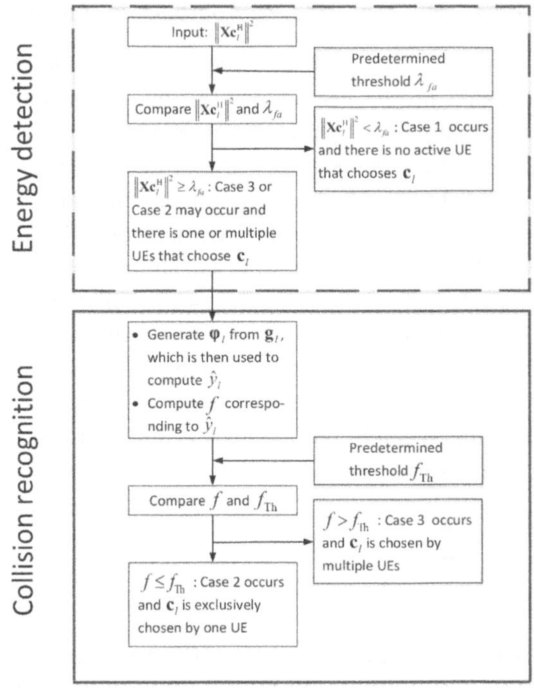

Figure 4.2 Flowchart for classifying Cases 1–3.

The statistic for energy detection is

$$T = \|\mathbf{x}\|^2. \tag{4.10}$$

When the required probability of false alarm P_{fa} is predetermined, the threshold λ_{fa} can be set as

$$\lambda_{\mathrm{fa}} = NN_0\left(M + \sqrt{2M}[Q]^{-1}(P_{\mathrm{fa}})\right). \tag{4.11}$$

By using the threshold λ_{fa}, Case 1 can be identified.

Since the statistic T can be approximated as a Gaussian distribution, the detection probability given by the Q function is

$$P_d = Q\left[\frac{Q^{-1}\left(P_{\mathrm{fa}}\right) - \sqrt{\frac{M}{2}\frac{NP_k\ell_k}{N_0}}}{1 + \frac{NP_k\ell_k}{N_0}}\right], \tag{4.12}$$

which implies that the detection is particularly accurate with a sufficiently large M. Then, given the preambles in use, the channel vectors are computed and buffered along with their associated preamble indices.

4.2.2 HOCs-based AMC Approach for Collision Recognition

Thanks to the channel hardening and favorable propagation characteristics of massive MIMO, estimation error due to preamble collision and interference can be rapidly decreased as the number of receive antennas grows [58], while the received collided signals processed at the BS can be considered as superposition coded signals. Moreover, to differentiate Cases 2 and 3, we need to additionally take account of the signal constellation. If quadrature amplitude modulation (QAM) is used, in Case 2, $\mathbf{y}_l$ will show 4 constellation points clearly. However, in Case 3, $\mathbf{y}_l$ will have more than 4 constellation points. Based on the above facts, a high-order cumulants (HOCs)-based automatic modulation classification (AMC) approach is adopted for classifying Cases 2 and 3.

We first describe the processed data packet in Eq. (4.8) of Cases 2 and 3. Assume channel vector $\mathbf{h}_k$ is estimated by a minimum mean square error (MMSE) detector. Then it can be given by [59]

$$\overline{\mathbf{h}}_k = \frac{\sqrt{P_k \ell_k}}{\sum_{j \in \{\zeta_s; k \in \zeta_s\}} P_j \ell_j + \frac{N_0}{N}} \mathbf{g}_k, \tag{4.13}$$

where ζ_s is the index set of the UEs that select the sth preamble. Due to the orthogonality property of the MMSE estimate, decomposition of the channel vector $\mathbf{h}_k$ can be generated as $\mathbf{h}_k = \overline{\mathbf{h}}_k + \hat{\boldsymbol{h}}_k$, where $\hat{\boldsymbol{h}}_k$ represents the uncorrelated estimation error following the distribution [52]

$$\hat{\boldsymbol{h}}_k \sim \mathcal{CN}\left(\mathbf{0}, \ell_k \left(1 - \frac{P_k \ell_k}{\sum_{j \in \{\zeta_s; k \in \zeta_s\}} P_j \ell_j + \frac{N_0}{N}}\right) \mathbf{I}_M\right). \tag{4.14}$$

Without loss of generality, we take the first active UE as an example to present the description, and mark the preamble picked by the first UE as the first preamble, so that the data sequence in Eq. (4.8) can be rewritten as

$$\begin{aligned} \mathbf{y}_1 = & \sum_{l \in \zeta_1} \boldsymbol{\varphi}_1 \sqrt{P_l} \overline{\mathbf{h}}_l \mathbf{d}_l + \sum_{l \in \zeta_1} \boldsymbol{\varphi}_1 \sqrt{P_l} \hat{\boldsymbol{h}}_l \mathbf{d}_l \\ & + \sum_{k \notin \zeta_1} \boldsymbol{\varphi}_1 \sqrt{P_k} \mathbf{h}_k \mathbf{d}_k + \boldsymbol{\varphi}_1 \mathbf{N}_2, \end{aligned} \tag{4.15}$$

where $\mathbf{y}_1$ associated with $|\zeta_1| = 1$ and $|\zeta_1| \geq 2$ corresponds to Cases 2 and 3, respectively. Compared to the first term which is viewed as the signal in Eq. (4.15), the estimation error, interference, and noise terms can be increasingly weaken as M grows.

Now, we present the HOCs-based AMC approach as CB is employed. We first compute the second-order moment of Eq. (4.15) when $D = 1$, which is given by

$$\begin{aligned}\mathbb{E}|y_1|^2 &= \sum_{l\in\zeta_1} P_l \mathbb{E}\left|\mathbf{g}_1^{\mathrm{H}}\bar{\mathbf{h}}_l\right|^2 + \sum_{l\in\zeta_1} P_l \mathbb{E}\left|\mathbf{g}_1^{\mathrm{H}}\hat{\boldsymbol{h}}_l\right|^2 \\ &+ \sum_{k\notin\zeta_1} P_k \mathbb{E}\left|\mathbf{g}_1^{\mathrm{H}}\mathbf{h}_k\right|^2 + \mathbb{E}\left|\mathbf{g}_1^{\mathrm{H}}\mathbf{N}_2\right|^2.\end{aligned} \tag{4.16}$$

Next, the signal is reorganized as $\hat{y}_1 = \left(\mathbf{g}_1^{\mathrm{H}}/|\mathbf{g}_1^{\mathrm{H}}|^2\right)\mathbf{Y} = y_1/|\mathbf{g}_1^{\mathrm{H}}|^2$. By utilizing the statistic in Eq. (4.16), we can further obtain that

$$\mathbb{E}|\hat{y}_1|^2 \xrightarrow{M\to\infty} \frac{\sum_{l\in\zeta_1} P_l^2 \ell_l^2}{\left(\sum_{l\in\zeta_1} P_l \ell_l + \frac{N_0}{N}\right)^2}. \tag{4.17}$$

Besides, the same statistic can be obtained with $\hat{y}_1 = y_1$ when ZFB is employed.

Note that $\mathbb{E}|\hat{y}_1|^2$ associated with $|\zeta_1| = 1$ approaches 1 as signal-to-noise ratio (SNR) grows while that associated with $|\zeta_1| \geq 2$ does not. The above fact can be utilized to distinguish Case 3 from Case 2. This can be attainable at a moderate SNR since the variance of noise is divided by the length of preamble sequences, N. Similarly, the approach can also be feasible in the case where more than two UEs collide. Moreover, in order to utilize the features of statistics and the constellation mentioned above to improve the classification performance, we then employ HOCs including the second-order cumulant C_{21}, the fourth-order cumulant C_{42}, the sixth-order cumulant C_{63}, and the eighth-order cumulant C_{80}, which can further mitigate the effect of noise. For the signal $\hat{y}_1$, the above HOCs can be expressed as $C_{21} = M_{21}$, $C_{42} = M_{42} - |M_{20}|^2 - 2M_{21}^2$, $C_{63} = M_{63} - 9M_{42}M_{21} + 9|M_{20}|^2 M_{21} + 12M_{21}^3$, and $C_{80} = M_{80} - 28M_{20}M_{60} - 35M_{40}^2 + 420M_{20}^2 M_{40} - 630M_{20}^4$, where $M_{pq} = \mathbb{E}\left[(\hat{y}_1)^{p-q}(\hat{y}_1^*)^q\right]$ [60]. Upon that, we define the feature used for classification as

$$f = |C_{80}|\, C_{63} C_{42} C_{21}. \tag{4.18}$$

Then, Cases 2 and 3 can be classified according to f via a predetermined threshold f_{Th}.

4.2.3 Data Decoding with SIC

Case 3 is referred to as preamble collision as multiple active UEs choose the same preamble. For example, let us consider the preamble picked by the first UE. Then, the probability of V that other UEs select the same preamble is

$$P_{\mathrm{sp}}(V) = B\left(V, K-1, \frac{1}{L}\right). \tag{4.19}$$

In practice, with $K = 20$ UEs out of 200 UEs active, i.e., each UE attempts to access the BS with a probability of 0.1 [61], and 64 available preambles, the probabilities of two UEs colliding, i.e., $V = 1$, and three UEs colliding, i.e., $V = 2$, are approximately 0.2236 and 0.0319, respectively. It can be foreseen that the cases of more UEs colliding can be negligible in terms of the overall probability. Therefore, for Case 3, we mainly consider two-UE and three-UE collisions, while the proposed approach used to detect the existence of three-UE collisions is outlined as follows.

First, all the signals associated with unique preambles are decoded and removed from $\mathbf{Y}$. That is, we find

$$\mathbf{Z} = \mathbf{Y} - \sum_{l \in \mathcal{A}} \mathbf{g}_l \mathbf{d}_l^{\mathrm{T}}, \tag{4.20}$$

where $\mathcal{A}$ is the set of the preambles with $|\zeta_l| = 1$ and $\mathbf{d}_l$ is the data packet transmitted by the active UE that chooses the lth preamble. Then, $\mathbf{R}_{\mathbf{Z}}$ only has the signals associated with $|\zeta_l| \geq 2$. In addition, we use $\mathcal{A}^*$ to denote the set of preambles with $|\zeta_l| \geq 2$.

By exploiting the channel hardening and favorable propagation characteristics and taking the limit $M \to \infty$, we have

$$\mathbf{R}_{\mathbf{Z}} = \frac{1}{D}\mathbf{Z}\mathbf{Z}^{\mathrm{H}} = \sum_{k \in \mathcal{B}} P_k \mathbf{h}_k \mathbf{h}_k^{\mathrm{H}} + N_0 \mathbf{I}, \tag{4.21}$$

where $\mathcal{B}$ is the index set of active UEs that experience preamble collisions. In general, the number of the active UEs that experience preamble collisions, i.e., $|\mathcal{B}|$, could be small if $L \gg K$.

Using the eigendecomposition of $\mathbf{R}_{\mathbf{Z}}$, $|\mathcal{B}|$ can be estimated. Then the existence of three-UE collisions can be generally discovered if $|\mathcal{B}| > 2|\mathcal{A}^*|$ holds, where $|\mathcal{A}^*|$ can be determined in the collision recognition phase. In addition, using an appropriate SIC approach that is applied to $\mathbf{Z}$, it is able to recover the signals transmitted from the active UEs that experience preamble collisions.

Here, we adopt the eigenvalue gap measure-based second-order statistic of the eigenvalues (SORTE) approach to determine the number of colliding UEs [62]. Computed from $\mathbf{R}_{\mathbf{Z}}$, the matrix of eigenvalues is given by

$$\mathbf{U} = \operatorname{diag}\left\{\sigma_1, \sigma_2, \dots, \sigma_M\right\}, \tag{4.22}$$

where σ_i is the ith eigenvalue. The SORTE approach chooses the value m as the estimation of $|\mathcal{B}|$, which minimizes the criterion

$$\mathrm{SORTE}(m) = \begin{cases} \dfrac{\mathrm{var}[\{\nabla\sigma_i\}_{i=m+1}^{M-1}]}{\mathrm{var}[\{\nabla\sigma_i\}_{i=m}^{M-1}]}, & \text{if } \mathrm{var}[\{\nabla\sigma_i\}_{i=m}^{M-1}] \neq 0, \\ +\infty, & \text{if } \mathrm{var}[\{\nabla\sigma_i\}_{i=m}^{M-1}] = 0, \end{cases} \tag{4.23}$$

where $m = 1, \ldots, M-1$, $\nabla\sigma_i = \sigma_i - \sigma_{i+1}$, $i = 1, \ldots, M-1$ and

$$\mathrm{var}\left[\{\nabla\sigma_i\}_{i=m}^{M-1}\right] = \frac{1}{M-m}\sum_{i=m}^{M-1}\left(\nabla\sigma_i - \frac{\sum_{i=m}^{M-1}\nabla\sigma_i}{M-m}\right)^2. \tag{4.24}$$

It is noteworthy that signal cancellation in Eq. (4.20) will additionally introduce interference due to the channel estimation error, which leads to a complicated noise environment. Since the ZFB outperforms the CB with the same receive antennas [63], the more favorable performance of SORTE can be obtained with ZFB.

4.3 Performance Analysis

We present the derivation of the success probability of the collision resolution protocol. The two-UE collision is studied, where the theoretic analysis consists of two parts, i.e., no collision and two-UE collisions. Without loss of generality, we take the first active UE as an example to derive the approximate analytic expressions. For simplicity, another UE that selects the same preamble as the first UE is referred to as the second UE.

Assuming that K UEs simultaneously carry out the RA procedure over the same channel, transmission of the first UE is viewed to be successful when its data packet is successfully decoded. Specifically, in the absence of collisions, the data packet of the first UE is considered to be successfully decoded when the signal-to-interference-plus-noise ratio (SINR) of its signal layer is greater than the threshold γ_{Th} [64]. Alternatively, in the presence of collisions, the data packet is considered to be successfully decoded with SIC when the SINR of the signal layer from the first UE (after stronger signal layers are detected and canceled) together with all the stronger signal layers is greater than the threshold γ_{Th}. In this section, the threshold γ_{Th} is determined by $\gamma_{\mathrm{Th}} = 2^{R_c} - 1$, where R_c is the code rate.

In the proposed RA procedure, given the length of preambles $N = 139$ and a sufficiently large M, the preamble detection probability can be nearly equal to 1. Besides, as shown in the simulation results, superimposed signals can be identified accurately even at a low P_t/N_0 by setting an appropriate threshold. Therefore, the probability of collision recognition can also approach 1. It is noteworthy that even though Case 2 can be falsely recognized as Case 3 at a low P_t/N_0, signals transmitted from the first UE as the strongest signal layer can still be decoded by SIC. As mentioned above, the success probability we focus on mainly depends on the data decoding phase, which is given by

$$P_{\mathrm{success}} = \sum_{V=0}^{K-1} P_{\mathrm{sp}}(V)\, P_{\mathrm{rc}}(V), \tag{4.25}$$

where V denotes the number of other co-channel UEs that select the same preamble with the first UE and $P_{rc}(V)$ represents the probability that the data transmission for the first UE is successful. Note that in the special case of two-UE collision, the probability reduces to

$$P_{2,\text{resolved}} = \left(1 - \tfrac{1}{L}\right)^{K-1} P_{rc}(0) + \tfrac{K-1}{L}\left(1 - \tfrac{1}{L}\right)^{K-2} P_{rc}(1). \tag{4.26}$$

1) **Success Probability with CB:** Then we calculate the success probability $P_{2,\text{resolved}}$ with CB. We first discuss the case associated with $\zeta_1 = \{1\}$, where no collision occurs, and derive the expression of $P_{rc,CB}(0)$. Let $P_{rc,CB}(0)$ be substituted by $P\left(\gamma_{CB}^0 > \gamma_{Th}\right)$, γ_{CB}^0 denotes the SINR of the received signal from the first UE, the superscript 0 indicates that there are no other UEs select the same preamble as the first UE, and the subscript CB represents that CB is applied. From Eq. (4.15), through substituting $\boldsymbol{\varphi}_1 = \mathbf{g}_1^H$, the SINR γ_{CB}^0 corresponding to the first UE is computed as

$$\gamma_{CB}^0 = \frac{P_1\left|\mathbf{g}_1^H\overline{\mathbf{h}}_1\right|^2}{P_1\mathbb{E}_{\hat{\boldsymbol{h}}_1}\left|\mathbf{g}_1^H\hat{\boldsymbol{h}}_1\right|^2 + \sum_{k=2}^{K} P_k\mathbb{E}_{\mathbf{v}_k}\left|\mathbf{g}_1^H\mathbf{h}_k\right|^2 + \|\mathbf{g}_1^H\|^2 N_0}. \tag{4.27}$$

Note that the interference in the denominator in Eq. (4.27) contains correlated terms since they have common path-loss terms as multipliers, while deriving exact expression of $P\left(\gamma_{CB}^0 > \gamma_{Th}\right)$ turns to be intractable. Therefore, channel estimation error $\hat{\boldsymbol{h}}_1$ and small-scale fading $\mathbf{v}_k$ are calculated with the help of expectation operators.

The SINR γ_{CB}^0 can be simplified as

$$\gamma_{CB}^0 = \frac{P_1^2\ell_1^2\left(M+1\right)}{\left(P_1\ell_1 + \frac{N_0}{N}\right)\sum_{k=2}^{K} P_k\ell_k + N_0\left(P_1\ell_1 + \frac{P_1\ell_1}{N} + \frac{N_0}{N}\right)}. \tag{4.28}$$

Derivation of Eq. (4.28): In the numerator of γ_{CB}^0, we have the signal power simplified as

$$P_1\left|\mathbf{g}_1^H\overline{\mathbf{h}}_1\right|^2 = P_1\left|\mathbf{g}_1^H\frac{\sqrt{P_1}\ell_1}{\sum_{l\in\zeta_1}P_l\ell_l + \frac{N_0}{N}}\mathbf{g}_1\right|^2 = P_1^2\ell_1^2\left|\tilde{\mathbf{g}}_1^H\tilde{\mathbf{g}}_1\right|^2, \tag{4.29}$$

where $\tilde{\mathbf{g}}_1$ is the normalized vector of which each entry has a unit variance. Furthermore, we denote $S_1 = \frac{1}{M}\left|\tilde{\mathbf{g}}_1^H\tilde{\mathbf{g}}_1\right|$ for convenience. Since S_1 can be rewritten as an average of M terms, where each term is the square of the independent sample of standard normal distribution, S_1 has the probability density function of Gamma distribution $\Gamma(y, M, 1/M)$. Then, with the help of the strong law of large numbers, we have ${S_1}^2 \xrightarrow{M\to\infty} 1 + 1/M$.

Next, we compute the terms in the denominator of Eq. (4.27) as

$$P_1\mathbb{E}\left|\mathbf{g}_1^{\mathrm{H}}\hat{\boldsymbol{h}}_1\right|^2 = P_1\mathbb{E}\left|\mathbf{g}_1^{\mathrm{H}}\right|^2\mathbb{E}\left|\hat{\boldsymbol{h}}_1\right|^2 = MP_1\ell_1\left(\sum_{l\in\zeta_1,l\neq 1} P_l\ell_l + \frac{N_0}{N}\right), \tag{4.30}$$

$$P_k\mathbb{E}_{\mathbf{v}_k}\left|\mathbf{g}_1^{\mathrm{H}}\mathbf{h}_k\right|^2 \approx P_k\ell_k\left(\sum_{l\in\zeta_1} P_l\ell_l + \frac{N_0}{N}\right)\left|\tilde{\mathbf{g}}_1^{\mathrm{H}}\mathbf{v}_k\right|^2. \tag{4.31}$$

From the central limit theorem, we have $I_k = \frac{1}{\sqrt{M}}\tilde{\mathbf{g}}_1^{\mathrm{H}}\mathbf{v}_k$, for $k \neq 1$, which converges to the standard normal distribution with a sufficiently large M. Therefore, $\left|I_k\right|^2$ converges to Gamma distribution $\Gamma(y,1,1)$. Then, $\left|I_k\right|^2$ is approximated with its expectation such that $\left|I_k\right|^2 = 1, k \neq 1$.

Lastly, the noise term in the denominator of Eq. (4.27) is simplified as

$$|\mathbf{g}_1^{\mathrm{H}}|^2 N_0 = M\left(\sum_{l\in\zeta_1} P_l\ell_l + \frac{N_0}{N}\right)N_0. \tag{4.32}$$

Then γ_{CB}^0 in Eq. (4.28) is computed by substituting Eqs. (4.29)–(4.32) into Eq. (4.27), which completes the derivation.

From [[52], Appendix B], the approximate probability of $P\left(\gamma_{\mathrm{CB}}^0 > \gamma_{\mathrm{Th}}\right)$ can be derived as

$$P\left(\gamma_{\mathrm{CB}}^0 > \gamma_{\mathrm{Th}}\right) \approx \sum_{n=1}^{N_c}\binom{N_c}{n}(-1)^{n+1}\int_0^{\infty} e^{-\frac{n\eta\gamma_{\mathrm{Th}}}{M+1}B_1(x)}B_3(x)\bar{f}_r(x)\,dx, \tag{4.33}$$

where N_c denotes the number of terms used in the calculation, $\eta = N_c\left(N_c!\right)^{-\frac{1}{N_c}}$, $\rho_t = \frac{P_t}{N_0}$, $B_1(x) = \frac{x^{\alpha(1-\epsilon)}}{P_tC^{1-\epsilon}}\left(1+\frac{1}{N}\right) + \frac{x^{2\alpha(1-\epsilon)}}{N\rho_t^2C^{2(1-\epsilon)}}$, $B_2(x) = x^{\alpha(1-\epsilon)} + \frac{x^{2\alpha(1-\epsilon)}}{N\rho_tC^{1-\epsilon}}$, and $B_3(x) = \left(\int_0^{\infty} e^{-\frac{n\eta\gamma_{\mathrm{Th}}}{M+1}B_2(x)s^{-\alpha(1-\epsilon)}}\bar{f}_r(s)\,ds\right)^{K-1}$.

Next, we consider the case of two-UE collision associated with $\zeta_1 = \{1,2\}$ and derive the probability $P_{\mathrm{rc,CB}}(1)$, which can be expressed as

$$P_{\mathrm{rc,CB}}(1) = P\left(\tilde{\gamma}_{\mathrm{UE1,CB}}^1 > \gamma_{\mathrm{Th}}\right) + P\left(\tilde{\gamma}_{\mathrm{UE2,CB}}^1 > \gamma_{\mathrm{Th}}, \hat{\gamma}_{\mathrm{UE1,CB}}^1 > \gamma_{\mathrm{Th}}\right), \tag{4.34}$$

where $P\left(\tilde{\gamma}_{\mathrm{UE1,CB}}^1 > \gamma_{\mathrm{Th}}\right)$ represents the probability that the signal transmitted from the first UE, as the strong signal layer, is decoded successfully. $\tilde{\gamma}_{\mathrm{UE1,CB}}^1$ denotes the SINR of the strong signal, where the subscript UE1 stands for the signal transmitted from the first UE is the strong signal, the subscript CB indicates that the CB is applied, and the superscript 1 shows that there is an arbitrary UE colliding with the first UE. Recall that the UE that chooses the same preamble as the first UE is referred to as the second UE. Then the SINR $\tilde{\gamma}_{\mathrm{UE1,CB}}^1$

of the first UE is given by

$$\tilde{\gamma}^1_{\mathrm{UE1,CB}} = \frac{P_1\left|\mathbf{g}_1^{\mathrm{H}}\overline{\mathbf{h}}_1\right|^2}{P_2\left|\mathbf{g}_1^{\mathrm{H}}\overline{\mathbf{h}}_2\right|^2 + \Delta_{\mathrm{CB}}}, \tag{4.35}$$

where $\Delta_{\mathrm{CB}} = P_1\mathbb{E}_{\hat{\boldsymbol{h}}_1}\left|\mathbf{g}_1^{\mathrm{H}}\hat{\boldsymbol{h}}_1\right|^2 + P_2\mathbb{E}_{\hat{\boldsymbol{h}}_2}\left|\mathbf{g}_1^{\mathrm{H}}\hat{\boldsymbol{h}}_2\right|^2 + \sum_{k=3}^{K} P_k\mathbb{E}_{\mathbf{v}_k}|\mathbf{g}_1^{\mathrm{H}}\mathbf{h}_k|^2 + |\mathbf{g}_1^{\mathrm{H}}|^2 N_0$. Conditioning on $R_1 = x$, $R_2 = y$, and $R_1 < R_2$, $P\left(\tilde{\gamma}^1_{\mathrm{UE1,CB}} > \gamma_{\mathrm{Th}}\right)$ can be rewritten as

$$\begin{aligned} P\left(\tilde{\gamma}^1_{\mathrm{UE1,CB}} > \gamma_{\mathrm{Th}}\right) &= P\left(\tilde{\beta}^1_{\mathrm{UE1,CB}} > \gamma_{\mathrm{Th}}\right) \\ &= \int_0^{\infty} dx \int_x^{\infty} P\left(\tilde{\beta}^1_{\mathrm{UE1,CB}} > \gamma_{\mathrm{Th}}|x,y\right)\bar{f}_r(x)\bar{f}_r(y)\,dy, \end{aligned} \tag{4.36}$$

where $\tilde{\beta}^1_{\mathrm{UE1,CB}} = (P_1\left|\mathbf{g}_1^{\mathrm{H}}\overline{\mathbf{h}}_1\right|^2 - P_2\left|\mathbf{g}_1^{\mathrm{H}}\overline{\mathbf{h}}_2\right|^2\gamma_{\mathrm{Th}})/\Delta_{\mathrm{CB}}$ and the calculation of $P\left(\tilde{\beta}^1_{\mathrm{UE1,CB}} > \gamma_{\mathrm{Th}}|x,y\right)$ follows the same line as in [[52], Appendix B].

Let $P\left(\tilde{\gamma}^1_{\mathrm{UE2,CB}} > \gamma_{\mathrm{Th}}, \hat{\gamma}^1_{\mathrm{UE1,CB}} > \gamma_{\mathrm{Th}}\right)$ represent the probability that the signal transmitted from the first UE, as the weak signal layer, is decoded successfully. Let $\tilde{\gamma}^1_{\mathrm{UE2,CB}}$ be the SINR of the strong signal, where the subscript UE2 denotes that the signal transmitted from the second UE is the strong signal, the subscript CB indicates that the CB is applied, and the superscript 1 means that there is another UE, i.e., the first UE, colliding with the second UE. Let $\hat{\gamma}^1_{\mathrm{UE1,CB}}$ be the SINR of the weak signal layer, where the subscript UE1 shows that the signal transmitted from the first UE is the weak signal. The SINR $\tilde{\gamma}^1_{\mathrm{UE2,CB}}$ of the second UE and the SINR $\hat{\gamma}^1_{\mathrm{UE1}}$ of the first UE are given by

$$\tilde{\gamma}^1_{\mathrm{UE2,CB}} = \frac{P_2\left|\mathbf{g}_1^{\mathrm{H}}\overline{\mathbf{h}}_2\right|^2}{P_1\left|\mathbf{g}_1^{\mathrm{H}}\overline{\mathbf{h}}_1\right|^2 + \Delta_{\mathrm{CB}}}, \tag{4.37}$$

$$\hat{\gamma}^1_{\mathrm{UE1,CB}} = \frac{P_1\left|\mathbf{g}_1^{\mathrm{H}}\overline{\mathbf{h}}_1\right|^2}{\Delta_{\mathrm{CB}}}, \tag{4.38}$$

respectively.

Similar to that in Eq. (4.36), the probability $P\left(\tilde{\gamma}^1_{\mathrm{UE2,CB}} > \gamma_{\mathrm{Th}}, \hat{\gamma}^1_{\mathrm{UE1,CB}} > \gamma_{\mathrm{Th}}\right)$ can be modified as $P\left(\tilde{\beta}^1_{\mathrm{UE2,CB}} > \gamma_{\mathrm{Th}}, \hat{\gamma}^1_{\mathrm{UE1,CB}} > \gamma_{\mathrm{Th}}\right)$, where $\tilde{\beta}^1_{\mathrm{UE2,CB}} = (P_2\left|\mathbf{g}_1^{\mathrm{H}}\overline{\mathbf{h}}_2\right|^2 - P_1\left|\mathbf{g}_1^{\mathrm{H}}\overline{\mathbf{h}}_1\right|^2\gamma_{\mathrm{Th}})/\Delta_{\mathrm{CB}}$. Then, $P\left(\tilde{\gamma}^1_{\mathrm{UE2,CB}} > \gamma_{\mathrm{Th}}, \hat{\gamma}^1_{\mathrm{UE1,CB}} > \gamma_{\mathrm{Th}}\right)$ is equivalent to $P\left(\min\{\tilde{\beta}^1_{\mathrm{UE2,CB}}, \hat{\gamma}^1_{\mathrm{UE1,CB}}\} > \gamma_{\mathrm{Th}}\right)$. Conditioning on $R_1 = x$, $R_2 = y$, and $R_1 > R_2$, given that $P_1\left|\mathbf{g}_1^{\mathrm{H}}\overline{\mathbf{h}}_1\right|^2$, $P_2\left|\mathbf{g}_1^{\mathrm{H}}\overline{\mathbf{h}}_2\right|^2$, and $P_1\left|\mathbf{g}_1^{\mathrm{H}}\overline{\mathbf{h}}_1\right|^2$ are calculated in Derivation of Eq. (4.28), we can obtain that

$$\min\left\{\tilde{\beta}^1_{\mathrm{UE2,CB}}, \hat{\gamma}^1_{\mathrm{UE1,CB}}\right\} = \begin{cases} \hat{\gamma}^1_{\mathrm{UE1,CB}}, & \text{if } 0 < R_2 < R_1\left(1+\gamma_{\mathrm{Th}}\right)^{-\frac{1}{2\alpha(1-\varepsilon)}} \\ \tilde{\beta}^1_{\mathrm{UE2,CB}}, & \text{if } R_1\left(1+\gamma_{\mathrm{Th}}\right)^{-\frac{1}{2\alpha(1-\varepsilon)}} < R_2 < R_1 \end{cases}. \tag{4.39}$$

As demonstrated above, $P\left(\tilde{\gamma}^1_{\mathrm{UE2,CB}} > \gamma_{\mathrm{Th}}, \hat{\gamma}^1_{\mathrm{UE1,CB}} > \gamma_{\mathrm{Th}}\right)$ can be decomposed as

$$\begin{aligned} &P\left(\tilde{\beta}^1_{\mathrm{UE2,CB}} > \gamma_{\mathrm{Th}}, \hat{\gamma}^1_{\mathrm{UE1,CB}} > \gamma_{\mathrm{Th}}\right) \\ &= \int_0^\infty dx \int_0^{x\left(1+\gamma_{\mathrm{Th}}\right)^{-\frac{1}{2\alpha(1-\varepsilon)}}} P\left(\hat{\gamma}^1_{\mathrm{UE1,CB}} > \gamma_{\mathrm{Th}} | x, y\right) \bar{f}_r(x) \bar{f}_r(y)\, dy \\ &+ \int_0^\infty dx \int_{x\left(1+\gamma_{\mathrm{Th}}\right)^{-\frac{1}{2\alpha(1-\varepsilon)}}}^{x} P\left(\tilde{\beta}^1_{\mathrm{UE2,CB}} > \gamma_{\mathrm{Th}} | x, y\right) \bar{f}_r(x) \bar{f}_r(y)\, dy, \end{aligned} \tag{4.40}$$

where the calculation of $P\left(\hat{\gamma}^1_{\mathrm{UE1,CB}} > \gamma_{\mathrm{Th}} | x, y\right)$ and $P\left(\tilde{\gamma}^1_{\mathrm{UE2,CB}} > \gamma_{\mathrm{Th}} | x, y\right)$ are also similar to that in [[52], Appendix B]. Finally, substituting Eqs. (4.36) and (4.40) into Eq. (4.34), the probability $P_{\mathrm{rc,CB}}(1)$ is obtained.

In the case of CB and no collision, the channel estimation error term, interference terms, and noise term decay at a rate of $\frac{1}{M+1}$ compared to the signal term. Therefore, the performance of CB can be near-optimal in massive MIMO system with a sufficiently large M in the case of no collision. However, in the presence of collisions, the performance of CB will heavily deteriorate due to the weak co-existing signal that becomes a prominent interference term since the weak signal term will not decay as M grows, as in Eqs. (4.35) and (4.37). Therefore, the performance of CB in the presence of collisions mainly depends on the signal power gap due to the different path losses besides the estimation error, interference, and noise.

2) **Success Probability with ZFB:** The success probability with ZFB can be different from that with CB. Since the vector utilized to decode data is computed from the whole channel estimation vectors, the collision among the other UEs is supposed to be considered. Let us calculate the success probability $P_{2,\mathrm{resolved}}$ in Eq. (4.26) with ZFB. We first discuss the case associated with $\zeta_1 = \{1\}$, where no collision occurs, and derive the expression of $P_{\mathrm{rc,ZF}}(0)$, which can be expressed as

$$P_{\mathrm{rc,ZF}}(0) = \sum_{S=1}^{\min\{L-1,K-1\}} \tilde{P}^0_{\mathrm{ZF}}(S)\, \tilde{P}\left(\gamma^0_{\mathrm{ZF}} > \gamma_{\mathrm{Th}} | S\right), \tag{4.41}$$

where the subscript ZF indicates that ZFB is employed.

Let $\tilde{P}_{\mathrm{ZF}}^{0}(S)$ represent the probability that S preambles, except the one picked by the first UE, are totally selected by the other $K-1$ UEs. Similar to that in [[61], Section III-B], $\tilde{P}_{\mathrm{ZF}}^{0}(S)$ is given by

$$\tilde{P}_{\mathrm{ZF}}^{0}(S)=\frac{\binom{L-1}{S}S!\left\{\begin{matrix}K-1\\S\end{matrix}\right\}}{(L-1)^{K-1}}, \tag{4.42}$$

where $\left\{\begin{matrix}K-1\\S\end{matrix}\right\}=(1/S!)\sum_{j=0}^{S}(-1)^{S-j}\binom{S}{j}j^{K-1}$ denotes the Stirling numbers of the second kind. Note that $\tilde{P}\left(\gamma_{\mathrm{ZF}}^{0}>\gamma_{\mathrm{Th}}|S\right)$ refers to the probability of $\gamma_{\mathrm{ZF}}^{0}>\gamma_{\mathrm{Th}}$ given S and K, where S preambles, except the one picked by the first UE, are totally selected by the other $K-1$ UEs and γ_{ZF}^{0} denotes the SINR corresponding to the first UE.

When ZFB is employed, $\mathbf{y}_1$ is the data sequence processed by the vector $\boldsymbol{\varphi}_1=\mathbf{a}_1$. Recall that $\mathbf{a}_1$ is the first column of $\left(\mathbf{A}^{\mathrm{H}}\mathbf{A}\right)^{-1}\mathbf{A}^{\mathrm{H}}$ and $\mathbf{A}$ is the channel estimation matrix defined as $\mathbf{A}=\left[\mathbf{g}_1,\mathbf{g}_2,\ldots,\mathbf{g}_{S+1}\right]\in\mathbb{C}^{M\times(S+1)}$. Besides, $\mathcal{S}=\{2,\ldots,S+1\}$ is the nonempty index set, where $s\in\mathcal{S}$ denotes the sth preamble selected by the UEs.

To simplify the derivation, the transformation from [[55], Section III-B] is applied to $\mathbf{y}_1$ in Eq. (4.15), and then the SINR γ_{ZF}^{0} can be expressed as

$$\gamma_{\mathrm{ZF}}^{0}=\frac{P_1\left|\mathbf{a}_1\bar{\mathbf{h}}_1\right|^2}{P_1\mathbb{E}_{\hat{\boldsymbol{h}}_1}\left|\mathbf{a}_1\hat{\boldsymbol{h}}_1\right|^2+\sum\limits_{j\in\{z_s;s\in\mathcal{S}\}}\mathbb{E}\left|\mathbf{a}_1\breve{\mathbf{h}}_j\right|^2+\sum\limits_{j\in\{\zeta_s\backslash z_s;s\in\mathcal{S}\}}P_k\mathbb{E}\left|\sqrt{2}\mathbf{a}_1\mathbf{h}_k\right|^2+\left|\mathbf{a}_1\right|^2N_0}, \tag{4.43}$$

where z_s is an arbitrary element in set ζ_s, $\zeta_s\backslash z_s$ represents the set ζ_s excluding the element z_s, and $\breve{\mathbf{h}}_{z_s}=\sum\limits_{j\in\zeta_s}P_j\mathbf{h}_j$.

Since the exact expression of γ_{ZF}^{0} is intractable to derive, channel estimation error $\hat{\boldsymbol{h}}_1$ and small-scale fading $\mathbf{v}_k$ are calculated with the help of expectation operators. The SINR γ_{ZF}^{0} can be simplified as

$$\gamma_{\mathrm{ZF}}^{0}=\frac{\frac{(P_1\ell_1)^2}{NP_1\ell_1+N_0}(M-S)}{\frac{P_1\ell_1N_0}{NP_1\ell_1+N_0}+N_0+\sum\limits_{k=2}^{K}\frac{M}{M-S}\frac{P_k\ell_kN_0}{NP_k\ell_k+N_0}+\sum\limits_{j\in\{\zeta_s\backslash z_s\}}P_j\ell_j}. \tag{4.44}$$

Derivation of Eq. (4.44): In the numerator of γ_{ZF}^{0}, since $\left|\mathbf{a}_1\mathbf{g}_1\right|^2=1$, we have the signal power simplified as

$$P_1\left|\mathbf{a}_1\bar{\mathbf{h}}_1\right|^2=\frac{P_1^2\ell_1^2}{\left(\sum_{l\in\zeta_1}P_l\ell_l+\frac{N_0}{N}\right)^2}. \tag{4.45}$$

From [54], $|\mathbf{a}_1|^{-2}$ follows the Chi-square distribution $\chi^2_{2(M-S)}$ and we have $|\mathbf{a}_1|^{-2} \xrightarrow{(M-S)\to\infty} (M-S)\left(\sum_{l\in\zeta_1} P_l \ell_l + N_0/N\right)$ via the strong law of large numbers. Approximating the values with their expectations as shown in [52], the terms $P_1\left|\mathbf{a}_1\hat{\boldsymbol{h}}_1\right|^2$ can be computed as

$$P_1\mathbb{E}\left|\mathbf{a}_1\hat{\boldsymbol{h}}_1\right|^2 \approx |\mathbf{a}_1|^2 P_1\ell_1\left(1-\frac{P_1\ell_1}{\sum_{l\in\zeta_1} P_l\ell_l+\frac{N_0}{N}}\right). \tag{4.46}$$

Then, we consider the second term in the denominator in Eq. (4.43), which can be approximated as its upper bound such that

$$\begin{aligned}\mathbb{E}\left|\mathbf{a}_1\breve{\mathbf{h}}_{\zeta_s}\right|^2 &= \frac{M}{M-S}\|\mathbf{a}_1\|^2\frac{\frac{N_0}{N}\sum_{l\in\zeta_s} P_l\ell_l}{\sum_{l\in\zeta_s} P_l\ell_l+\frac{N_0}{N}}\\ &\lessapprox \frac{M}{M-S}\|\mathbf{a}_1\|^2\sum_{l\in\zeta_s}\frac{\frac{N_0}{N}P_l\ell_l}{P_l\ell_l+\frac{N_0}{N}}.\end{aligned} \tag{4.47}$$

From [55], the second term is approximated as

$$P_i\mathbb{E}_{\mathbf{v}_i}\left|\sqrt{2}\mathbf{a}_1\mathbf{h}_i\right|^2 = |\mathbf{a}_1|^2 P_i\ell_i\mathbb{E}_{\mathbf{v}_i}\left|\frac{\sqrt{2}\mathbf{a}_1\mathbf{v}_i}{|\mathbf{a}_1|}\right|^2 \approx |\mathbf{a}_1|^2 P_i\ell_i. \tag{4.48}$$

Finally, γ^0_{ZF} in Eq. (4.44) is obtained by substituting Eqs. (4.45)–(4.48) into Eq. (4.43), which completes the derivation.

As mentioned in above, given S and K, similar to that with CB, the approximate probability of $P\left(\gamma^0_{\mathrm{ZF}} > \gamma_{\mathrm{Th}}|S\right)$ can be calculated as

$$\begin{aligned}P\left(\gamma^0_{\mathrm{ZF}} > \gamma_{\mathrm{Th}}|S\right) \approx &\sum_{n=1}^{N_c}\binom{N_c}{n}(-1)^{n+1}\\ &\int_0^\infty e^{-\frac{n\eta\gamma_{\mathrm{Th}}}{M-S}S_1(x)}S_3(x)S_4(x)\bar{f}_r(x)\,dx,\end{aligned} \tag{4.49}$$

where $S_1(x) = \frac{(N+1)x^{\alpha(1-\varepsilon)}}{N\rho_t C^{1-\varepsilon}} + \frac{x^{2\alpha(1-\varepsilon)}}{N\rho_t^2 C^{2(1-\varepsilon)}}$, $S_2(x) = x^{\alpha(1-\varepsilon)} + \frac{x^{2\alpha(1-\varepsilon)}}{N\rho_t C^{1-\varepsilon}}$, $S_3(x) = \left(\int_0^\infty e^{-\frac{n\gamma_{\mathrm{Th}}\eta M}{M-S}S_2(x)\frac{u^{-\alpha(1-\varepsilon)}}{N\rho_t C^{1-\varepsilon}u^{-\alpha(1-\varepsilon)}+1}}\bar{f}_r(u)\,du\right)^S$, $S_4(x) = \left(\int_0^\infty e^{-\frac{n\gamma_{\mathrm{Th}}\eta}{M-S}S_2(x)S_5(v)}\bar{f}_r(v)\,dv\right)^{K-1-S}$, and $S_5(v) = v^{-\alpha(1-\varepsilon)}\left(\frac{1}{N\rho_t C^{1-\varepsilon}v^{-\alpha(1-\varepsilon)}+1}+1\right)$.

Next, we consider the case of two-UE collision associated with $\zeta_1 = \{1,2\}$, and derive the probability $P_{\mathrm{rc,ZF}}(1)$ which can be expressed as

$$\begin{aligned}P_{\mathrm{rc,ZF}}(1) = &\sum_{S=1}^{\min\{L-1,K-2\}}\tilde{P}^1_{\mathrm{ZF}}(S)\left[\tilde{P}\left(\beta^1_{\mathrm{UE1,ZF}} > \gamma_{\mathrm{Th}}|S\right)\right.\\ &\left.+P\left(\tilde{\beta}^1_{\mathrm{UE2,ZF}} > \gamma_{\mathrm{Th}}, \hat{\gamma}^1_{\mathrm{UE1,ZF}} > \gamma_{\mathrm{Th}}|S\right)\right].\end{aligned} \tag{4.50}$$

Here, $\tilde{P}_{\mathrm{ZF}}^{1}(S)$ represent the probability that S preambles, except the one picked by the first UE, are totally selected by the other $K-2$ UEs, which is given by

$$\tilde{P}_{\mathrm{ZF}}^{1}(S)=\frac{\binom{L-1}{S}S!\left\{ {K-2 \atop S} \right\}}{(L-1)^{K-2}}, \tag{4.51}$$

where $\left\{ {K-2 \atop S} \right\}=(1/S!)\sum_{j=0}^{S}(-1)^{S-j}\binom{S}{j}j^{K-2}$ denotes the Stirling numbers of the second kind.

Similar to that with CB, $\tilde{\beta}_{\mathrm{UE1,ZF}}^{1}$, $\tilde{\beta}_{\mathrm{UE2,ZF}}^{1}$, and $\hat{\gamma}_{\mathrm{UE1,ZF}}^{1}$ with ZFB can be rewritten as

$$\tilde{\beta}_{\mathrm{UE1,ZF}}^{1}=\frac{P_1\left|\mathbf{a}_1\bar{\mathbf{h}}_1\right|^2-P_2\left|\mathbf{a}_1\bar{\mathbf{h}}_2\right|^2\gamma_{\mathrm{Th}}}{\Delta_{\mathrm{ZF}}}, \tag{4.52}$$

$$\tilde{\beta}_{\mathrm{UE2,ZF}}^{1}=\frac{P_2\left|\mathbf{a}_1\bar{\mathbf{h}}_2\right|^2-P_1\left|\mathbf{a}_1\bar{\mathbf{h}}_1\right|^2\gamma_{\mathrm{Th}}}{\Delta_{\mathrm{ZF}}}, \tag{4.53}$$

$$\hat{\gamma}_{\mathrm{UE1,ZF}}^{1}=\frac{P_1\left|\mathbf{a}_1\bar{\mathbf{h}}_1\right|^2}{\Delta_{\mathrm{ZF}}}, \tag{4.54}$$

respectively, where $\Delta_{\mathrm{ZF}}=P_1\mathbb{E}_{\hat{\boldsymbol{h}}_1}\left|\mathbf{a}_1\hat{\boldsymbol{h}}_1\right|^2+P_2\mathbb{E}_{\hat{\boldsymbol{h}}_2}\left|\mathbf{a}_1\hat{\boldsymbol{h}}_2\right|^2+\sum_{j\in\{z_s;s\in S\}}\mathbb{E}\left|\mathbf{a}_1\breve{\mathbf{h}}_j\right|^2+\sum_{j\in\{\zeta_s\backslash z_s;s\in S\}}P_k\mathbb{E}\left|\sqrt{2}\mathbf{a}_1\mathbf{h}_k\right|^2+\left|\mathbf{a}_1\mathbf{N}_2\right|^2$. Note that the calculations of the probabilities $P\left(\tilde{\beta}_{\mathrm{UE1,ZF}}^{1}>\gamma_{\mathrm{Th}}|S\right)$ and $P\left(\tilde{\beta}_{\mathrm{UE2,ZF}}^{1}>\gamma_{\mathrm{Th}},\hat{\gamma}_{\mathrm{UE1,ZF}}^{1}>\gamma_{\mathrm{Th}}|S\right)$ follow the same line as in Eqs. (4.36) and (4.40), respectively. As a result, by substituting the above calculations into Eq. (4.50), the probability $P_{\mathrm{rc,ZF}}(1)$ is obtained.

Similar to CB, in the absence of collision, the channel estimation error term, interference terms, and noise term decay at a rate of $\frac{1}{M-S}$ compared to the signal term. Since $M \gg S$ can generally hold in massive MIMO systems and ZF can have insignificant interference terms, the performance of ZF outperforms that of CB. In the presence of collisions, the weak co-existing signal that becomes a prominent interference term cannot be further suppressed by employing ZF. However, since other interference terms after ZF become insignificant compared to CB, the performance of ZF in the presence of collisions is preferable.

4.4 Simulation Results

We present numerical results to evaluate the collision resolution protocol, where each UE is assumed to be randomly located in the cell and its distance from the BS follows a Rayleigh distribution with mean $0.5\sqrt{1/\lambda}$.

In practice, although the total number of UEs in a cell can be large for massive access scenarios, the number of UEs that attempt to access the network may not be large in a certain time interval, as the sparse user activity is considered in most IoT and MTC applications. Thus, for example, given $\lambda = 3/\text{k m}^2$, it is assumed that there are 200 UEs in the cell and each UE is active with a probability of 0.05, i.e., $K = 10$, within a given time interval. The assumption is applied in the simulations. Besides, since the number of active UEs may not always remain constant at different uplink transmission time intervals, the impact of various K, i.e., $K = 2 - 30$, on performance is taken into consideration.

For the first active UE, let us recall that its transmission power $P_1 = P_t(\ell_1)^{-\varepsilon}$ consists of open loop power P_t and FPC factor $P_{\text{FPC}} = \ell_1^{-\varepsilon}$ derived from the measurement of the system information, where $\ell_1 = C(R_1)^{-\alpha}$ and the constant term C is set to be 1 for simplicity. Assuming that each UE has the capability to implement the FPC, we focus on the open loop transmission signal-to-noise ratio (OSNR), i.e., P_t/N_0. It is noteworthy that the arrived SNR at the BS is lower than OSNR. Particularly, simulation parameters are given in Table 4.1.

Prior to evaluate the success probability of the RA protocol, we present the results of the HOCs-based approach for collision recognition. Feature f designed in Eq. (4.18) as a function of OSNR with $M = 400$ and $\epsilon = 0.7$ is illustrated in Fig. 4.3. As we can see, the feature f of the signal with a two-UE collision approaches 0, while that of the signal without a collision tends to converge on a much lower value as OSNR grows. Based on the facts above, the cases of preamble collisions can be recognized through a predetermined threshold. However, due to the randomness of the transmission power of UEs, the threshold should be set to a relatively low value to deal with some extreme cases, i.e., the transmission power of the weak signal is much smaller than that of the strong signal, where the feature of a signal with a collision may become similar to that without a collision.

Figure 4.4 presents the probability of correct estimation of $|\zeta_1|$ versus OSNR with $M = 400$ and $\epsilon = 0.7$, where $|\zeta_1| = 1$ and $|\zeta_1| = 2$ correspond to Cases 2 and

Table 4.1 Parameters Setting.

Modulation	4QAM
Number of simultaneous UEs K	2~ -30 UEs
Number of antennas M	100, 400, 800
Number of orthogonal preambles L	64
Length of preamble sequences N	139
Length of data packets D	60
Path loss exponent α	4
BS density λ	$3/\text{k m}^2$
SINR threshold γ_{Th}	$-2, -1, 0$ dB

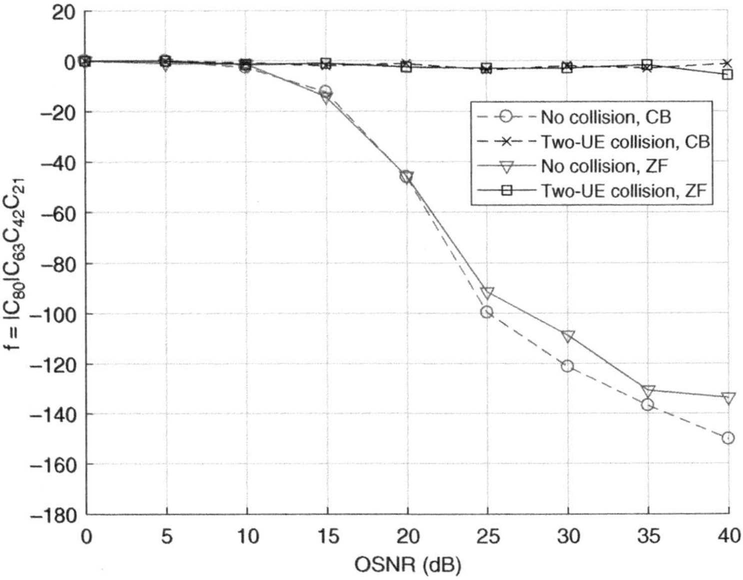

Figure 4.3 Feature f versus OSNR with $M = 400$ and $\epsilon = 0.7$.

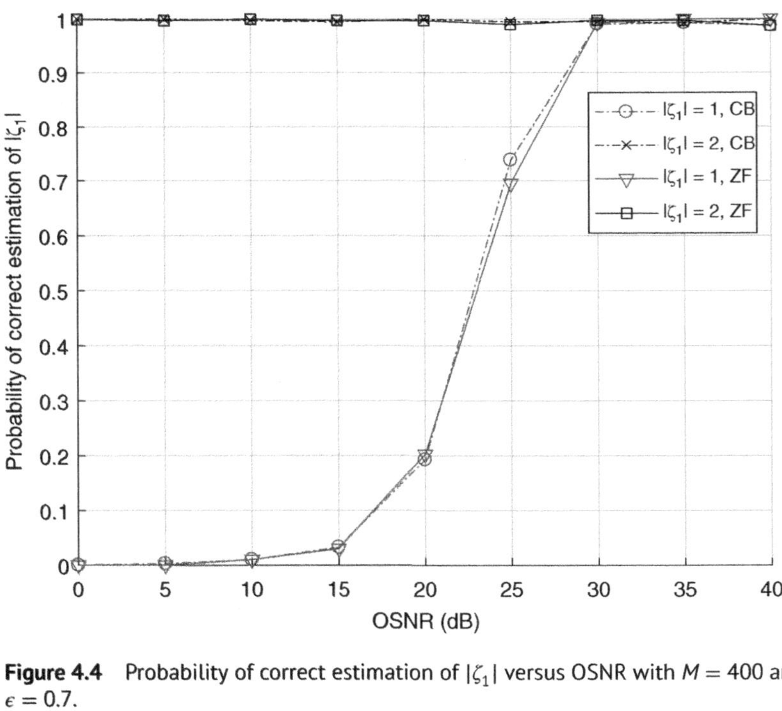

Figure 4.4 Probability of correct estimation of $|\zeta_1|$ versus OSNR with $M = 400$ and $\epsilon = 0.7$.

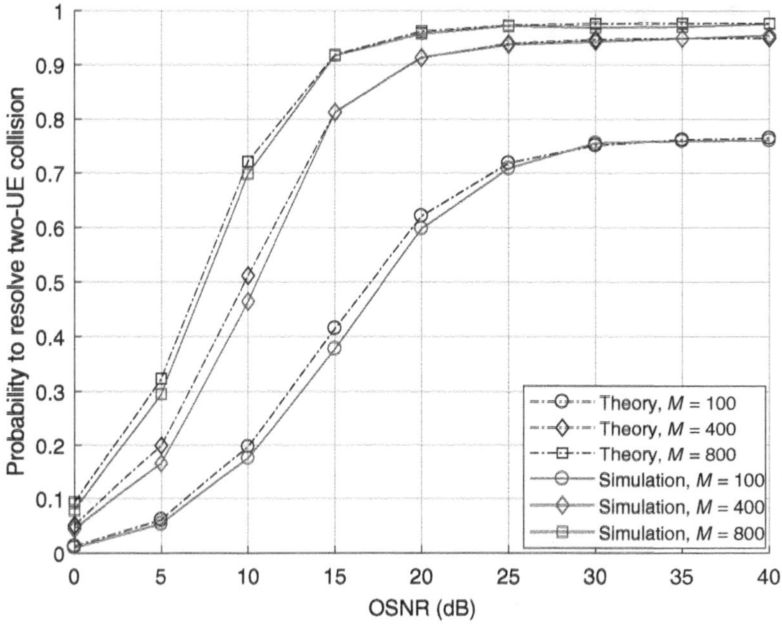

Figure 4.5 Probability $P_{rc,CB}(1)$ of resolving a two-UE collision versus OSNR with CB, various M, $\gamma_{Th} = 0$ dB, $\epsilon = 0.7$, and $K = 10$.

3, respectively, while the threshold to decide the two cases is set to be $f_{Th} = -80$. As shown in the figure, Case 2 is easy to be recognized as Case 3 over the range of OSNR ≤ 25 dB, which means that the noise is considered as a useful signal. However, the noise can be distinguished due to its different frame structure, and the data from the first UE is still able to be successfully decoded. Besides, the probability of correctly identifying Case 2 gradually increases to 1 as OSNR grows. Furthermore, Case 3 can be efficiently identified by the AMC approach at a broad range of OSNR.

1) **Performance on resolving a two-UE collision with CB:** The probabilities of resolving a two-UE collision as a function of OSNR with CB under different values of M are presented in Fig. 4.5. The observation shows that the analytic and simulation results match well with each other. Besides, we can see that the probabilities tend to get saturated as OSNR increases, in which the performance primarily depends on M. As M grows, the probabilities can reach the saturation point at a lower OSNR for better performance. Thus, the analysis can be verified that the impact of estimation error, interference, and noise decays as M increases.

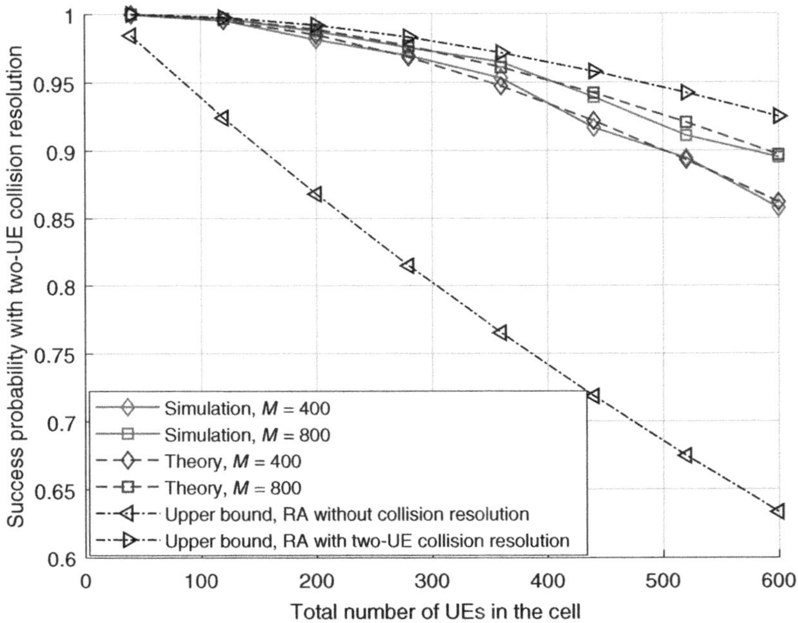

Figure 4.6 Success probability $P_{2,\mathrm{resolved}}$ versus the total number of UEs in the cell with CB, $\gamma_{\mathrm{Th}} = 0$ dB, OSNR = 30 dB, and $\epsilon = 0.7$.

Success probabilities as a function of the total number of UEs in the cell with CB are presented in Fig. 4.6. Recall that the success probability in the case of two-UE collision is given in Eq. (4.26). If the two-UE collision is perfectly resolved, we can obtain $P_{\mathrm{rc}}(0) = 1$ and $P_{\mathrm{rc}}(1) = 1$. Therefore, the upper bound of RA with two-UE collision resolution is given by $\left(1 - \frac{1}{L}\right)^{K-1} + \frac{K-1}{L}\left(1 - \frac{1}{L}\right)^{K-2}$. However, if the collision resolution is not employed, the first UE can complete RA successfully only when there are no other UEs that select the same preamble, i.e., $P_{\mathrm{rc}}(V) = 0$, $V \geq 1$. Thus, the upper bound of RA without collision resolution is given by $\left(1 - \frac{1}{L}\right)^{K-1}$. The observation shows that the proposed protocol can effectively improve the success probability of RA when the number of available preambles is limited. As K becomes larger, we can see that the success probability decreases gradually, while the performance can be improved as M increases.

Focusing on the impact of FPC, Fig. 4.7 illustrates the probability $P_{\mathrm{rc,CB}}(1)$ as a function of ϵ with adequate open loop transmission power $P_t = 80$ dB and $M = 400$. As shown in the figure, the probability becomes nearly optimal over the range of $0.7 \leq \epsilon \leq 0.85$ and drops for $\epsilon < 0.7$ or $\epsilon > 0.85$. When ϵ is small,

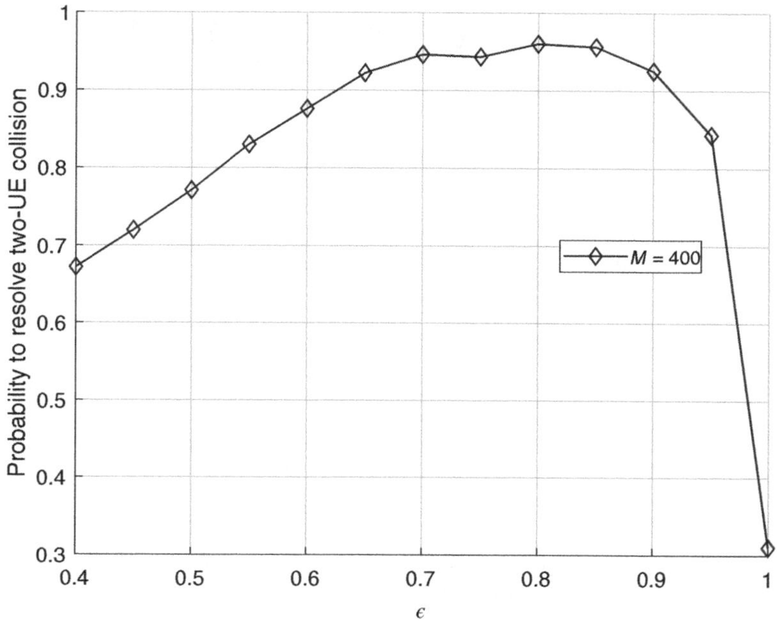

Figure 4.7 Probability $P_{rc,CB}(1)$ of resolving a two-UE collision versus ϵ with CB, $M = 400$, OSNR $= 80$ dB, and $\gamma_{Th} = 0$ dB.

the signals received at the BS become weak, which can be easily drowned in the co-channel interference. Thus, when the two UEs collide, the strong one can be successfully recognized and decoded while the weak one may not. Alternatively, when ϵ approaches 1, since the difference between the channel gains of the two colliding UEs reduces, the decreased SINR of the strong signal from the superimposed signals may lead to the unsuccessful SIC.

2) **Performance on resolving a two-UE collision with ZFB:** In Fig. 4.8, the probabilities of resolving a two-UE collision as a function of OSNR with ZFB are illustrated. As we can see, the analytic results coincide with the simulation ones, in which the success probabilities are asymptotically saturated as OSNR grows. Moreover, compared with the results of CB, ZFB can achieve the same performance by utilizing fewer receive antennas, which reveals that ZFB is more effective in suppressing the co-channel interference than CB.

Success probabilities as a function of the total number of UEs in the cell with ZFB are presented in Fig. 4.9. Similar to the observation in Fig. 4.6, the success probability is effectively improved relative to the RA without collision resolution. As M becomes larger, we can see that the success probability approaches

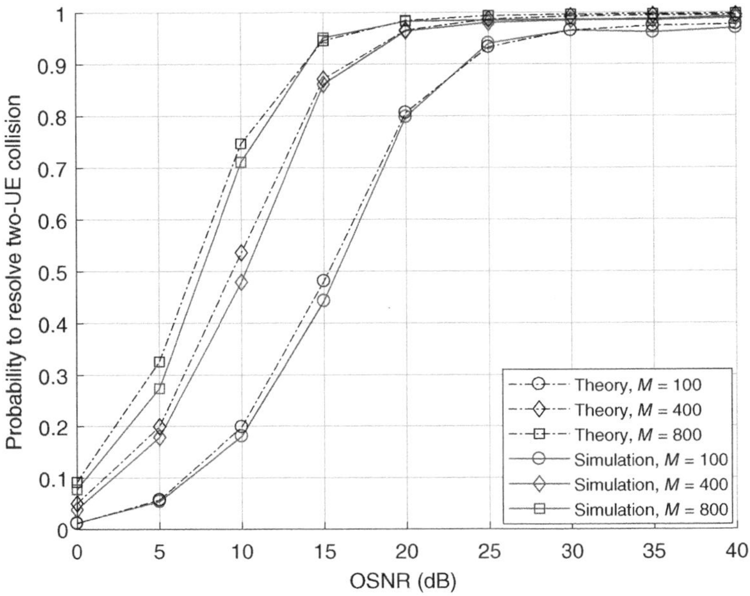

Figure 4.8 Probability $P_{rc,ZF}(1)$ of resolving a two-UE collision versus OSNR with ZFB, various M, $\gamma_{Th} = 0\,$dB, $\epsilon = 0.7$, and $K = 10$.

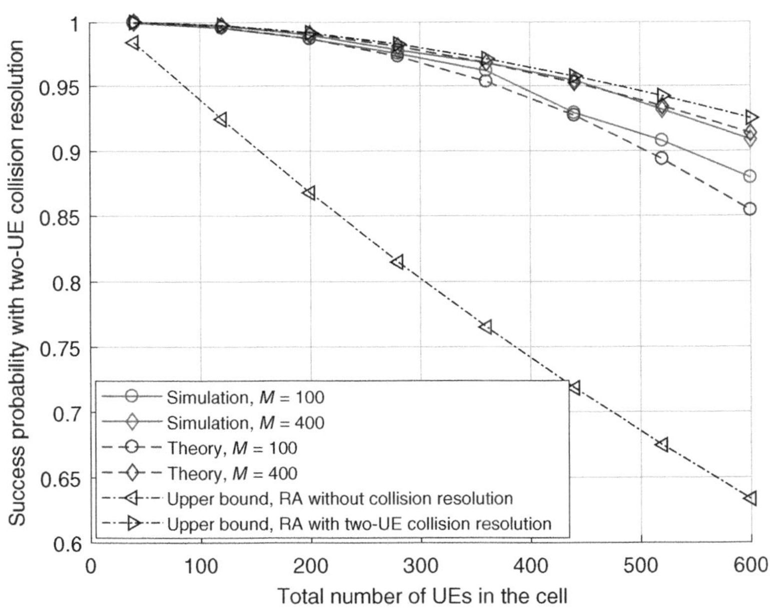

Figure 4.9 Success probability $P_{2,resolved}$ versus the total number of UEs in the cell with ZFB, $\gamma_{Th} = 0\,$dB, OSNR $= 30\,$dB, and $\epsilon = 0.7$.

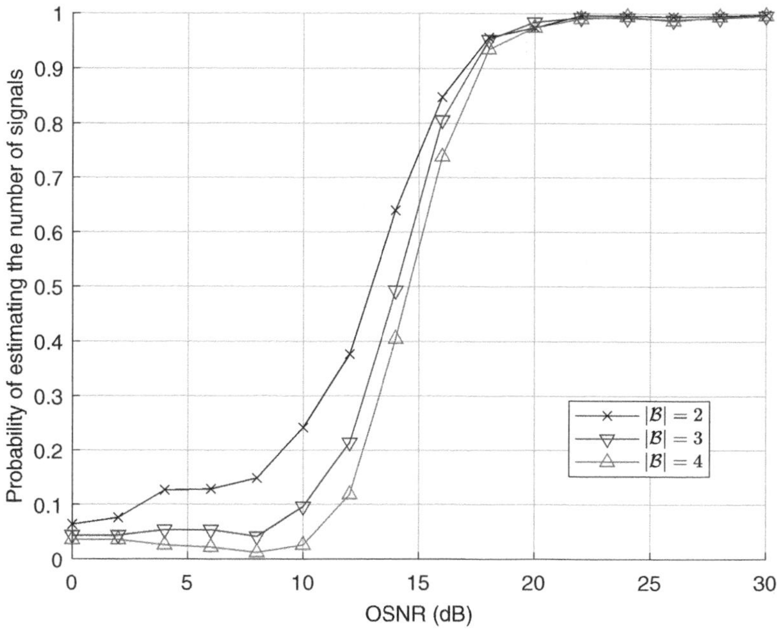

Figure 4.10 Probability of correctly estimating $|\mathcal{B}|$ via SORTE versus OSNR with ZFB, $M = 400$, $\epsilon = 0.7$, $K = 10$, and $D = 60$.

the upper bound of RA with collision resolution. It is clear that ZFB is more effective than CB in accommodating more UEs.

Figure 4.10 shows the probability of correctly estimating the number of collided signals via SORTE approach as a function of OSNR. The results show that based on SORTE method, $|\mathcal{B}|$ can be accurately estimated over the range of OSNR > 20 dB. Hence, the existence of three-UE collisions can be effectively discovered by employing the proposed approach.

Figure 4.11 plots the achievable probabilities of resolving a three-UE collision with ZFB as a function of OSNR. In this case, we assume the BS always attempts to decode the third signal layer, if the two-UE collision has been discovered. Besides, the cases considering SORTE may cause performance degradation at a relatively low OSNR, due to the incorrect estimation of $|\mathcal{B}|$. As the figure shows, the success probability gets saturated at around 80% with $\gamma_{\text{Th}} = 0\,\text{dB}$, while the performance can be slightly improved when M increases from 400 to 800. For the received signal with a three-UE collision, i.e., the signal superimposed by three collided signals after ZFB detection, the SINR of the superimposed signal to be canceled should be sufficiently high, which may

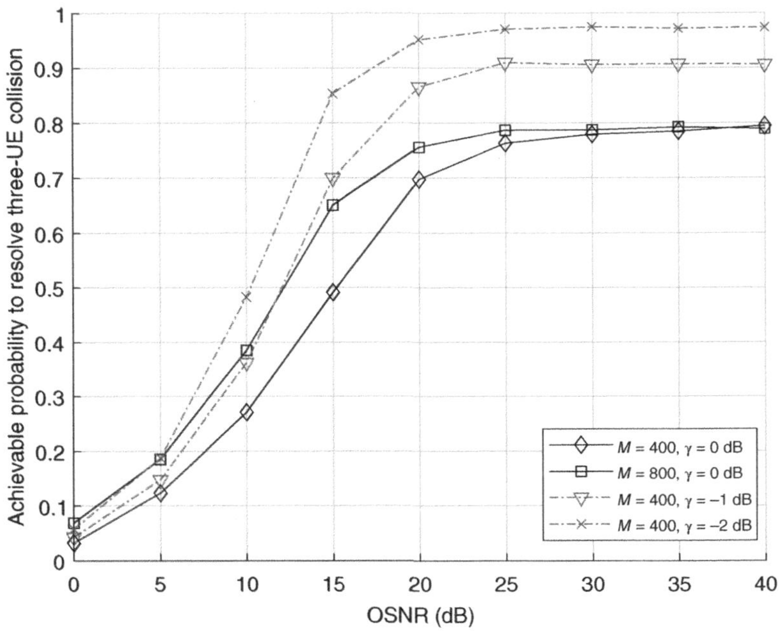

Figure 4.11 Achievable probability of resolving a three-UE collision versus OSNR with $\epsilon = 0.7$ and $K = 10$.

not be practical as there is no coordination among the UEs to optimize their distances from the BS. However, the probability can still be improved at the expense of sacrificing the code rate R_c, i.e., decreasing γ_{Th}.

4.5 Conclusion and Remarks

In this chapter, we introduced a massive MIMO-based grant-free RA with a collision resolution protocol to support massive access for MTC communications in IoT networks. To resolve preamble collisions, a HOCs-based AMC approach was studied to effectively recognize the collision cases. By exploiting the channel hardening and favorable propagation characteristics of massive MIMO, collided signals can be first processed as superposition coded signals by employing the beamforming, and then were successfully decoded using the SIC. In addition, theoretical analyses on the success probability of the collision resolution with CB and ZFB were studied.

For collision resolution, there are some other issues for future work. For example, the path loss compensation fraction ϵ can be optimized to maximize the success probability and the analysis to see the impact of inter-cell interference on the success probability is another meaningful extension. Besides, in order to significantly improve the performance of collided signal detection, we will introduce some important parameter estimation methods in Chapter 5.

5

Multiple Delay Estimation for Collided Signals

- In machine-type communication (MTC), active devices randomly choose resource blocks (RB) and access to an access point (AP) [65, 66]. The arrival times of multiple signals can be different as devices are randomly located in a cell. Although multiple devices may choose the same RB, their signals can be detected by taking into account the difference between the arrival times of multiple signals, which is called non-orthogonal random access (NORA) [67]. NORA utilizes the different arrival times to identify multiple devices with the identical preamble, while the successive interference cancellation (SIC) can be employed to decode multiple signals when they coexist in the same RB [68–70]. Thus, the detection of the event of collision via the estimation of round-trip delay (RTD) in NORA becomes important. In this chapter, we introduce different approaches to estimate of RTD of multiple signals and determine the number of multiple signals [71, 72].

5.1 System Model

Consider an random access (RA) system consisting of an AP and multiple devices in a single-cell, with radius R, and assume that devices are randomly located in the cell.

We consider an orthogonal frequency division multiplexing (OFDM) system where subcarriers are divided into multiple groups. In particular, we assume that there are B RBs and each RB consists of $L + L_{cp}$ subcarriers (thus, the total number of subcarriers is $B(L + L_{cp})$), where L is the length of the pilot signal and L_{cp} is the length of cyclic prefix (CP), which is used to avoid the inter-symbol interference (ISI). A device with data packets, which is called an active device, sends a pilot signal of length L in a randomly chosen block among B RBs, and then the AP is able to estimate the channel state information (CSI) for coherent detection of the following data transmissions.

IoT Signal Detection, First Edition. Rui Han, Jingjing Wang, Lin Bai, and Jianwei Liu.

In MTC, the RTD varies from one device to another. The maximum RTD, denoted by τ_{mrtd}, is proportional to the cell radius R. That is,

$$\tau_{\mathrm{mrtd}} = \frac{2R}{c},$$

where c is the speed of light. In addition, denote by τ_{mds} the maximum delay spread and by T_s the sampling interval. Then, the maximum length of the effective channel impulse response (CIR), which includes the delay spread and RTD, is given by

$$P = \left\lceil \frac{\tau_{\mathrm{mrtd}} + \tau_{\mathrm{mds}}}{T_s} \right\rceil,$$

where $P \leq L_{\mathrm{cp}}$. For example, if the cell radius is 1 km, the maximum RTD becomes 6.67 μs. If the maximum delay spread is 5 μs (which is a value according to the extended typical urban (ETU) model in [73]) and $T_s = 1\,\mu\mathrm{s}$ (for a bandwidth of 1 MHz), P becomes 12.

Next, we consider the detection of collision based on RTD estimation, we mainly focus on one RB for RA, saying RB 1. In addition, considering the transmission efficiency, we assume that $L_{\mathrm{cp}} = P$. Suppose that there are K active devices that choose RB 1. For $K > 1$, within the RB, the interference is inevitable and we have NORA among K active devices. Denote by d_k the RTD of active device k (or the arrival time of the transmitted signal by active device k) in the discrete time, i.e., after sampling, where $d_k \in \{0, 1, \dots, d_{\max}\}$. Here, d_k is quantized in time[1] and $d_{\max}$ is the maximum RTD in the discrete time. In addition, Q denotes the maximum tap number of active device k's CIR due to delay spread in the discrete time, and assume that

$$d_{\max} = P - Q.$$

In Fig. 5.1, we consider a scenario of an AP and multiple devices in an RA system. For example, MTC devices for the industrial manufacturing monitoring can be considered. Suppose that there are two active devices. The effective CIRs with different RTDs are illustrated.

Let $\mathbf{s} = [s_0 \; \dots \; s_{L-1}]^{\mathrm{T}}$ denote the pilot sequence of length L transmitted by a device to allow the AP to estimate its CSI. Suppose that the AP has M receive antennas. Then, the received signal at the AP (through RB 1) through the mth antenna in the frequency-domain after removing CP becomes

$$\mathbf{y}_m = \sum_{k=1}^{K} \mathbf{H}_{m,k}\mathbf{s} + \mathbf{n}_m, \; m = 1, \dots, M, \tag{5.1}$$

where $\mathbf{H}_{m,k} = \mathrm{diag}(H_{m,k,0}, \dots, H_{m,k,L-1})$ is the (frequency-domain) diagonal channel matrix from device k to the mth antenna at the AP and $\mathbf{n}_m \sim \mathcal{CN}(\mathbf{0}, N_0\mathbf{I})$ is the

1 For example, d_k is given by $d_k = \lfloor \frac{\tau_k}{T_s} \rfloor$, where τ_k is the actual RTD of device k.

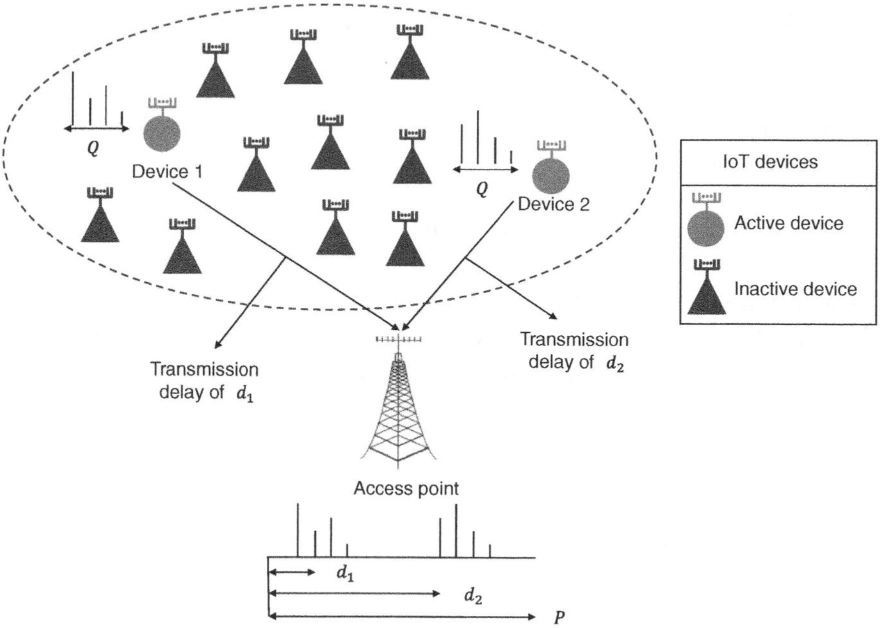

Figure 5.1 System model of two active devices with different RTDs.

background noise. Here, the channel coefficient corresponding to the lth subcarrier is given by

$$H_{m,k,l} = \sum_{p=0}^{P-1} c_{m,k,p} e^{-\frac{j2\pi pl}{L}},$$

where $\{c_{m,k,p}\}$ is the effective CIR of device k (to the mth antenna at the AP), and $l = 0, \ldots, L-1$. For convenience, let $\mathbf{S} = \text{diag}(s_0, \ldots, s_{L-1})$. Then, $\mathbf{y}_m$ can also be expressed as

$$\mathbf{y}_m = \mathbf{S} \sum_{k=1}^{K} \mathbf{h}_{m,k} + \mathbf{n}_m, \tag{5.2}$$

where $\mathbf{h}_{m,k} = [H_{m,k,0} \ \cdots \ H_{m,k,L-1}]^{\text{T}}$. In order to detect collision, the AP is to determine the value of K from $\{\mathbf{y}_m\}$ with known $\mathbf{S}$.

Unfortunately, the collision detection approaches in the frequency-domain may not provide a reliable performance, because the devices can have different RTDs. That is, for an active device with a long RTD, $c_{m,k,0}$ becomes 0, which means that the signal power from this active device is not taken into account by the correlator-based detector whose output is $\mathbf{s}^{\text{H}}\mathbf{y}_m$ unless the device's RTD is 0. To avoid this problem, the approach in the time-domain can be considered as in [74].

Let $\mathbf{c}_{m,k} = [c_{m,k,0} \cdots c_{m,k,P-1}]^{\mathrm{T}}$. Then, $\mathbf{h}_{m,k}$ can be expressed as

$$\mathbf{h}_{m,k} = \mathbf{F}_P \mathbf{c}_{m,k}, \tag{5.3}$$

where $[\mathbf{F}_P]_{l,p} = e^{-\frac{j2\pi pl}{L}}$ for $l = 0, \dots, L-1$ and $p = 0, \dots, P-1$, and the received signal becomes

$$\mathbf{y}_m = \mathbf{S}\mathbf{F}_P \sum_{k=1}^{K} \mathbf{c}_{m,k} + \mathbf{n}_m = \mathbf{\Psi}\mathbf{c}_m + \mathbf{n}_m, \tag{5.4}$$

where $\mathbf{\Psi} = \mathbf{S}\mathbf{F}_P$ and $\mathbf{c}_m = \sum_{k=1}^{K} \mathbf{c}_{m,k}$, which is referred to as the composite channel vector. The arrival time of the transmitted signal by active device k is d_k, and let $\mathbf{v}_{m,k}$ denote the CIR of length Q without RTD from device k to the mth antenna at the AP. Then, $\mathbf{c}_{m,k}$ can be given by

$$\mathbf{c}_{m,k} = [\underbrace{0 \dots 0}_{d_k} \mathbf{v}_{m,k}^{\mathrm{T}} \underbrace{0 \dots 0}_{P-Q-d_k}]^{\mathrm{T}}. \tag{5.5}$$

For convenience, let $\mathbf{v}_{m,k} = [v_{m,k,0} \cdots v_{m,k,Q-1}]^{\mathrm{T}}$, i.e., $v_{m,k,q}$ denotes the qth element of $\mathbf{v}_{m,k}$. From Eq. (5.4), we have

$$\mathbf{z}_m = \mathbf{\Psi}^{\dagger}\mathbf{y}_m = \mathbf{c}_m + \mathbf{\Psi}^{\dagger}\mathbf{n}_m, \tag{5.6}$$

where $\mathbf{\Psi}^{\dagger}$ is the pseudo-inverse of $\mathbf{\Psi}$, which is given by $\mathbf{\Psi}^{\dagger} = (\mathbf{\Psi}^{\mathrm{H}}\mathbf{\Psi})^{-1}\mathbf{\Psi}^{\mathrm{H}}$.

Then, we can consider an optimal estimation of RTDs, $\{d_k\}$, based on the maximum likelihood (ML) approach.

5.2 Multiple Delay Estimation

5.2.1 ML Detection Algorithm

In the following, we focus on the ML approach to obtain the estimates of RTDs, which can be used for the determination of the number of active devices or signals, K, under the assumptions below:

A1) For $\mathbf{v}_{m,k}$, multipath Rayleigh fading channels are considered, i.e.,

$$v_{m,k,q} \sim \mathcal{CN}(0, \sigma_{k,q}^2),\ q = 0, \dots, Q-1, \tag{5.7}$$

where $\sigma_{k,q}^2$ is the variance of $v_{m,k,q}$. Furthermore, the $v_{m,k,q}$'s are independent.

A2) Since the RTD, d_k, is proportional to the distance between the AP and active device k, $\sigma_{k,q}^2$ is given by

$$\sigma_{k,q}^2 = \nu(d_k)^{-\eta}\bar{\sigma}_q^2,$$

where $\{\overline{\sigma}_q^2\}$ is the normalized power delay profile (PDP), η is the path loss exponent, and $\nu(d)$ is the distance function of the RTD, d, i.e., $\nu(d)$ represents the distance from an MTC device to the AP when the RTD of the MTC device is d.

From Eq. (5.4), the conditional probability density function of $\{\mathbf{y}_m\}$ for given $\{\mathbf{c}_{m,k}\}$ can be written as

$$f(\{\mathbf{y}_m\} \mid \{\mathbf{c}_{m,k}\}) = C_1 \exp\left(-\frac{1}{N_0} \sum_{m=1}^{M} \|\mathbf{y}_m - \mathbf{\Psi}\mathbf{c}_m\|^2 \right), \tag{5.8}$$

where C_1 is the normalization constant. From Eq. (5.5), we can see that $\mathbf{c}_k$ is determined by $\{d_k, \mathbf{v}_k\}$. In addition, according to **A1** and **A2**, we have

$$\begin{aligned} f(\mathbf{c}_{1,k}, \dots, \mathbf{c}_{M,k} \mid d_k) &= C_2 \left(\prod_{q=0}^{Q-1} \sigma_{k,q}^2 \right)^{-M} \times \exp\left(-\sum_{m=1}^{M} \sum_{q=0}^{Q-1} \frac{|c_{m,k,d_k+q}|^2}{\nu(d_k)^{-\eta} \overline{\sigma}_q^2} \right) \\ &= C_2 \left(\prod_{q=0}^{Q-1} \sigma_{k,q}^2 \right)^{-M} \times \exp\left(-\sum_{m=1}^{M} \mathbf{c}_{m,k} \mathbf{R}_k^{\dagger} \mathbf{c}_{m,k} \right), \end{aligned} \tag{5.9}$$

where C_2 is constant, $\mathbf{R}_k$ is given by

$$\mathbf{R}_k = \text{diag}(\underbrace{0, \dots, 0}_{d_k}, \sigma_{k,0}^2, \dots, \sigma_{k,Q-1}^2, \underbrace{0, \dots, 0}_{P-Q-d_k}),$$

and $\mathbf{R}_k^{\dagger}$ is the pseudo-inverse of $\mathbf{R}_k$, which is given by

$$\mathbf{R}_k^{\dagger} = \text{diag}(\underbrace{0, \dots, 0}_{d_k}, \frac{1}{\sigma_{k,0}^2}, \dots, \frac{1}{\sigma_{k,Q-1}^2}, \underbrace{0, \dots, 0}_{P-Q-d_k}). \tag{5.10}$$

Note that $\mathbf{R}_k$ depends on d_k. Then, since the $\mathbf{c}_{m,k}$'s are independent (for given $\{d_1, \dots, d_K\}$) under **A1**, the likelihood function of $\{d_1, \dots, d_K\}$ becomes

$$f(\{\mathbf{y}_m\} \mid d_1, \dots, d_K) = \prod_{m=1}^{M} f(\mathbf{y}_m \mid d_1, \dots, d_K), \tag{5.11}$$

where

$$\begin{aligned} f(\mathbf{y}_m \mid d_1, \dots, d_K) = \int \cdots \int & f(\mathbf{y}_m \mid \mathbf{c}_{m,1}, \dots, \mathbf{c}_{m,K}) \\ & \times \prod_k f(\mathbf{c}_{m,k} \mid d_k) d\mathbf{c}_{m,1} \cdots d\mathbf{c}_{m,K}. \end{aligned} \tag{5.12}$$

To derive $f(\mathbf{y}_m \mid d_1, \dots, d_K)$, we define the diagonal matrix as

$$\mathbf{D} = \begin{bmatrix} \mathbf{R}_1^{\dagger} & \cdots & \mathbf{0} \\ \vdots & \ddots & \vdots \\ \mathbf{0} & \cdots & \mathbf{R}_K^{\dagger} \end{bmatrix}. \tag{5.13}$$

In addition, define

$$\Delta(d_1, \dots, d_K) = \prod_{k=1}^{K} \prod_{q=0}^{Q-1} \sigma_{k,q}^{2M}. \tag{5.14}$$

Note that $\Delta(d_1, \dots, d_K)$ depends on $\{d_k\}$ according to **A2**.

Lemma 5.1 *Let*

$$\mathbf{W}^{\mathrm{H}}\mathbf{W} = \mathbf{D} + \frac{1}{N_0}\mathbf{B}^{\mathrm{H}}\mathbf{B}, \tag{5.15}$$

where $\mathbf{B} = \mathbf{1}_{1,K} \bigotimes \boldsymbol{\Psi} = [\boldsymbol{\Psi} \cdots \boldsymbol{\Psi}]$. *Then, we have*

$$f(\mathbf{y}_m \,|\, d_1, \dots, d_K) \propto \frac{\exp\left(\mathbf{y}_m^{\mathrm{H}}\mathbf{B}(\mathbf{W}^{\mathrm{H}}\mathbf{W})^{-1}\mathbf{B}^{\mathrm{H}}\mathbf{y}_m\right)}{\Delta(d_1, \dots, d_K)}. \tag{5.16}$$

Proof: For convenience, we omit the antenna index m. It can be shown that

$$\begin{aligned} f(\mathbf{y} \,|\, \mathbf{c}_1, \dots, \mathbf{c}_K)\prod_k f(\mathbf{c}_k \,|\, d_k) &\propto \frac{\exp\left(-\frac{1}{N_0}\|\mathbf{y} - \boldsymbol{\Psi}\sum_k \mathbf{c}_k\|^2 - \sum_k \mathbf{c}_k^{\mathrm{H}}\mathbf{R}_k^{\dagger}\mathbf{c}_k\right)}{\Delta(d_1, \dots, d_k)} \\ &= \frac{\exp\left(-\frac{1}{N_0}\|\mathbf{y} - \mathbf{B}\mathbf{c}\|^2 - \mathbf{c}^{\mathrm{H}}\mathbf{D}\mathbf{c}\right)}{\Delta(d_1, \dots, d_K)} \\ &= e^{-\|\mathbf{W}\mathbf{c} - \mathbf{b}\|^2 - \frac{1}{N_0}\|\mathbf{y}\|^2} \frac{e^{\|\mathbf{b}\|^2}}{\Delta(d_1, \dots, d_K)}, \end{aligned}$$

where $\mathbf{c} = [\mathbf{c}_1^{\mathrm{T}} \cdots \mathbf{c}_K^{\mathrm{T}}]^{\mathrm{T}}$ and $\mathbf{b} = \mathbf{W}^{-\mathrm{H}}\mathbf{B}^{\mathrm{H}}\mathbf{y}$. Then, we have

$$\int \cdots \int f(\mathbf{y} \,|\, \mathbf{c}_1, \dots, \mathbf{c}_K)\prod_k f(\mathbf{c}_k \,|\, d_k) d\mathbf{c}_1 \cdots d\mathbf{c}_K \propto \frac{\exp\left(\|\mathbf{b}\|^2\right)}{\Delta(d_1, \dots, d_K)}, \tag{5.17}$$

which completes the proof.

According to Lemma 5.1, the ML estimation of RTDs can be formulated as

$$\begin{aligned} \underset{d_1,\dots,d_K}{\operatorname{argmax}} \prod_{m=1}^{M} f\left(\mathbf{y}_m \middle| d_1, \dots, d_K\right) = \underset{d_1,\dots,d_K}{\operatorname{argmax}} \sum_{m=1}^{M} \mathbf{y}_m^{\mathrm{H}}\mathbf{B}\left(\mathbf{D} + \frac{1}{N_0}\mathbf{B}^{\mathrm{H}}\mathbf{B}\right)^{-1} \mathbf{B}^{\mathrm{H}}\mathbf{y}_m \\ - \ln \Delta(d_1, \dots, d_K), \end{aligned} \tag{5.18}$$

where $\mathbf{D}$ is a matrix that is a function of the RTDs, $d_1, \dots, d_K$, as in Eqs. (5.10) and (5.13). An exhaustive search can be considered to find the ML estimates of the RTDs in Eq. (5.18). In this case, since $d_k \in \{0, 1, \dots, P - Q\}$, the complexity order becomes $(P - Q + 1)^K$. Thus, if K is large, the complexity can be high.

5.2.2 CAVI Detection Algorithm

As a method of machine learning for approximating probability densities, variational inference (VI) is widely used to approximate posterior densities for Bayesian models. The principle of VI is to perform optimization rather than sampling, where a family of densities is first posited, and then a member of the family that is close to the target density can be found. Therefore, we apply VI to the ML estimation problem in Eq. (5.18) in order to obtain approximate solutions with low computational complexity.

We assume that the RTDs, $\{d_k\}$, are independent random variables, which are seen as the latent variables. Let $\rho_k(d) = \Pr(d_k = d)$, which is referred to as the variational distribution, and let $\rho(\mathbf{d}) = \prod_{k=1}^{K} \rho_k(d_k)$, where $\mathbf{d} = [d_1 \; \dots \; d_K]^{\mathrm{T}}$. In VI [25, 75], the estimation of $\mathbf{d}$ becomes the optimization problem formulated as

$$\rho^*(\mathbf{d}) = \underset{\rho(\mathbf{d}) \in \mathcal{P}}{\operatorname{argmin}} \, \mathrm{KL}(\rho(\mathbf{d}) \| \Pr(\mathbf{d} \mid \{\mathbf{y}_m\})), \tag{5.19}$$

where $\mathcal{P}$ represents the collection of all possible distributions of $\mathbf{d}$, $\Pr(\mathbf{d} \mid \{\mathbf{y}_m\})$ is the posteriori probability of $\mathbf{d}$, and $\mathrm{KL}(\cdot)$ is the Kullback–Leibler (KL) divergence defined as

$$\mathrm{KL}(\rho(\mathbf{d}) \| p(\mathbf{d})) = \mathbb{E}_\rho \left[\ln \frac{\rho(\mathbf{d})}{p(\mathbf{d})} \right] = \sum_{\mathbf{d}} \rho(\mathbf{d}) \ln \frac{\rho(\mathbf{d})}{p(\mathbf{d})}.$$

Then, from $\rho^*(\mathbf{d}) = \prod_k \rho_k^*(d_k)$, the estimate of d_k from $\rho^*(\mathbf{d})$ can be found as

$$\hat{d}_k = \underset{d \in \{0, \dots, d_{\max}\}}{\operatorname{argmax}} \; \rho_k^*(d). \tag{5.20}$$

The minimization of KL divergence in Eq. (5.19) is equivalent to the maximization of the evidence lower bound (ELBO) [25], which is given by

$$\mathrm{ELBO}(\rho) = \mathbb{E}[\ln f(\{\mathbf{y}_m\}, \mathbf{d})] - \mathbb{E}[\ln \rho(\mathbf{d})], \tag{5.21}$$

where the expectation is carried out over $\mathbf{d}$. Let $\mathbf{d}_{-k} = [d_1 \; \dots \; d_{k-1} \, d_{k+1} \; \dots \; d_K]^{\mathrm{T}}$. Then, for given $\mathbf{d}_{-k}$, it can be shown that

$$\rho_k^*(d_k) \propto \exp\left(\mathbb{E}_{-k}[\ln f(d_k \mid \mathbf{d}_{-k}, \{\mathbf{y}_m\})]\right), \tag{5.22}$$

where the expectation, denoted by $\mathbb{E}_{-k}$, is carried out over $\mathbf{d}_{-k}$. The coordinate ascent variational inference (CAVI) algorithm [25, 43] is to update $\rho_k(d)$, where $k = 1, \dots, K$ and $d \in \{0, \dots, d_{\max}\}$, with fixing other variational distributions, $\rho_m(d)$, where $m \neq k$, through iterations.

In Eq. (5.22), the term on the right-hand side (RHS) can be further shown as

$$\begin{aligned} &\mathbb{E}_{-k}[\ln f(d_k \mid \mathbf{d}_{-k}, \{\mathbf{y}_m\})] \\ &\quad = \mathbb{E}_{-k}[\ln f(\{\mathbf{y}_m\} \mid \mathbf{d}) \Pr(d_k)] - \mathbb{E}_{-k}[\ln f(\{\mathbf{y}_m\} \mid \mathbf{d}_{-k})] \\ &\quad = \mathbb{E}_{-k}[\ln f(\{\mathbf{y}_m\} \mid \mathbf{d})] + \ln \Pr(d_k) + \text{const.} \end{aligned} \tag{5.23}$$

Note that if d_k is equally likely, $\ln \Pr(d_k)$ becomes a constant term. Since $\ln f(\{\mathbf{y}_m\} \mid \mathbf{d}) = \sum_m \ln f(\mathbf{y}_m \mid \mathbf{d})$, it can be shown that

$$\mathbb{E}_{-k}[\ln f(\{\mathbf{y}_m\} \mid \mathbf{d})] = \sum_m \mathbb{E}_{-k}[\ln f(\mathbf{y}_m \mid \mathbf{d})]. \tag{5.24}$$

Thus, for the CAVI algorithm, we need to find a closed-form expression for $\mathbb{E}_{-k}[\ln f(\mathbf{y}_m \mid \mathbf{d})]$. To this end, we can find an expression of $\ln f(\mathbf{y}_m \mid \mathbf{d})$ as follows.

Lemma 5.2 *Suppose that the s_l's are constant-modulus signals, i.e., $|s_l|$ is same for all l. Let $\mathbf{u}_m = \mathbf{\Psi}^{\mathrm{H}}\mathbf{y}_m = \mathbf{F}_P^{\mathrm{H}}\mathbf{S}^{\mathrm{H}}\mathbf{y}_m$, $m = 1, \dots, M$. In addition, we define*

$$r_p(d) = \begin{cases} \nu(d)^{-\eta}\bar{\sigma}_{p-d}^2, & \text{if } p \in \{d, \dots, d+Q-1\}, \\ 0, & \text{o.w.}, \end{cases} \tag{5.25}$$

which is also the pth diagonal element of $\mathbf{R}_k$ if $d = d_k$. Then, we have

$$\ln f(\mathbf{y}_m \mid \mathbf{d}) = N_0 \sum_p |u_{m,p}|^2 \beta_p(\mathbf{d}) + MQ\eta \sum_k \ln \nu(d_k) + \text{const.}, \tag{5.26}$$

where

$$\beta_p(\mathbf{d}) = 1 - \frac{\alpha N_0}{\alpha N_0 + \sum_k r_p(d_k)}. \tag{5.27}$$

Here, $\alpha = \frac{1}{|s_l|^2 L} = \frac{1}{\|\mathbf{s}\|^2}$ (for constant-modulus signals).

Proof: From Eq. (5.16), we have

$$\begin{aligned} \ln f(\mathbf{y}_m \mid \mathbf{d}) &= \mathbf{y}_m^{\mathrm{H}}\mathbf{B}\left(\mathbf{D} + \frac{1}{N_0}\mathbf{B}^{\mathrm{H}}\mathbf{B}\right)^{-1}\mathbf{B}^{\mathrm{H}}\mathbf{y}_m \\ &\quad - \ln \Delta(d_1, \dots, d_K) + \text{const.} \end{aligned} \tag{5.28}$$

From Eq. (5.14) and according to **A2**, we can show that

$$\begin{aligned} \ln \Delta(d_1, \dots, d_K) &= M \sum_{k=1}^{K} \sum_{q=0}^{Q-1} \ln(\nu(d_k)^{-\eta}\bar{\sigma}_q^2) \\ &= -QM\eta \sum_{k=1}^{K} \ln \nu(d_k) + \text{const.}, \end{aligned} \tag{5.29}$$

where the constant term is the term that does not depend on $\{d_k\}$.

Using the matrix inversion lemma, it can be shown that

$$\begin{aligned} \mathbf{B}\left(\mathbf{D} + \frac{1}{N_0}\mathbf{B}^{\mathrm{H}}\mathbf{B}\right)^{-1}\mathbf{B}^{\mathrm{H}} &= \mathbf{\Omega} - \mathbf{\Omega}(N_0\mathbf{I} + \mathbf{\Omega})^{-1}\mathbf{\Omega} \\ &= N_0\mathbf{\Omega}(N_0\mathbf{I} + \mathbf{\Omega})^{-1}, \end{aligned} \tag{5.30}$$

where

$$\mathbf{\Omega} = [\mathbf{\Psi} \cdots \mathbf{\Psi}]\mathbf{D}^{\dagger} \begin{bmatrix} \mathbf{\Psi}^{\mathrm{H}} \\ \vdots \\ \mathbf{\Psi}^{\mathrm{H}} \end{bmatrix} = \sum_k \mathbf{\Psi}\mathbf{R}_k\mathbf{\Psi}^{\mathrm{H}}. \tag{5.31}$$

Since $\mathbf{I} = \alpha \mathbf{S}\mathbf{F}\mathbf{F}^{\mathrm{H}}\mathbf{S}^{\mathrm{H}}$, it can be further shown that

$$\begin{aligned}(N_0\mathbf{I} + \boldsymbol{\Omega})^{-1} &= \left(N_0\mathbf{I} + \sum_k \boldsymbol{\Psi}\mathbf{R}_k\boldsymbol{\Psi}^{\mathrm{H}}\right)^{-1} \\ &= \left(\mathbf{S}\mathbf{F}\begin{bmatrix} N_0\alpha\mathbf{I} + \sum_k \mathbf{R}_k & \mathbf{0} \\ \mathbf{0} & N_0\alpha\mathbf{I}\end{bmatrix}\mathbf{F}^{\mathrm{H}}\mathbf{S}^{\mathrm{H}}\right)^{-1} \\ &= (\mathbf{S}\mathbf{F})^{-\mathrm{H}}\begin{bmatrix} N_0\alpha\mathbf{I} + \sum_k \mathbf{R}_k & \mathbf{0} \\ \mathbf{0} & N_0\alpha\mathbf{I}\end{bmatrix}^{-1}(\mathbf{S}\mathbf{F})^{-1}.\end{aligned}$$

From this, after some manipulations, we can show that

$$\begin{aligned}N_0\boldsymbol{\Omega}(N_0\mathbf{I} + \boldsymbol{\Omega})^{-1} &= N_0\mathbf{S}\mathbf{F}_P\left(\sum_k \mathbf{R}_k\right)\left(N_0\alpha\mathbf{I} + \sum_k \mathbf{R}_k\right)^{-1}\mathbf{F}_P^{\mathrm{H}}\mathbf{S}^{\mathrm{H}} \\ &= N_0\boldsymbol{\Psi}\left(\sum_k \mathbf{R}_k\right)\left(N_0\alpha\mathbf{I} + \sum_k \mathbf{R}_k\right)^{-1}\boldsymbol{\Psi}^{\mathrm{H}}. \end{aligned} \tag{5.32}$$

Substituting Eq. (5.32) into Eq. (5.30), and then into Eq. (5.28), we have

$$\begin{aligned}\mathbf{y}_m^{\mathrm{H}}\mathbf{B}\left(\mathbf{D} + \frac{1}{N_0}\mathbf{B}^{\mathrm{H}}\mathbf{B}\right)^{-1}\mathbf{B}^{\mathrm{H}}\mathbf{y}_m &= N_0\mathbf{u}_m^{\mathrm{H}}\left(\sum_k \mathbf{R}_k\right)\left(N_0\alpha\mathbf{I} + \sum_k \mathbf{R}_k\right)^{-1}\mathbf{u}_m \\ &= N_0\sum_p |u_{m,p}|^2\frac{\sum_k r_p(d_k)}{N_0\alpha + \sum_k r_p(d_k)} \\ &= N_0\sum_p |u_{m,p}|^2\left(1 - \frac{N_0\alpha}{N_0\alpha + \sum_k r_p(d_k)}\right),\end{aligned} \tag{5.33}$$

where the second equality is due to the fact that $\mathbf{R}_k$ is diagonal. Substituting Eqs. (5.29) and (5.33) into Eq. (5.28), we can obtain Eq. (5.26), which completes the proof.

For given variational distributions of $\rho_m(d)$, where $m \neq k$, we need to carry out the expectation $\mathbb{E}_{-k}[\ln f(\mathbf{y}_m \mid \mathbf{d})]$. From Eqs. (5.23), (5.24), and (5.26), if d_k is equally likely, we have

$$\begin{aligned}&\mathbb{E}_{-k}[\ln f(d_k \mid \mathbf{d}_{-k}, \{\mathbf{y}_m\})] \\ &= \sum_{m=1}^{M}\mathbb{E}_{-k}[\ln f(\mathbf{y}_m \mid \mathbf{d})] + \text{const.} \\ &= N_0\sum_p\left(\sum_m |u_{m,p}|^2\right)\mathbb{E}_{-k}\left[\beta_p(\mathbf{d})\right] + MQ\eta\left(\ln \nu(d_k) + \sum_{m\neq k}\mathbb{E}[\ln \nu(d_m)]\right) + \text{const.}\end{aligned} \tag{5.34}$$

Unfortunately, although the d_k's are assumed to be independent, it is troublesome to obtain $\mathbb{E}_{-k}\left[\beta_p(\mathbf{d})\right]$. Thus, we resort to an approximation as

$$\begin{aligned}
\mathbb{E}_{-k}\left[\beta_p(\mathbf{d})\right] &= 1 - \mathbb{E}_{-k}\left[\frac{N_0\alpha}{N_0\alpha + r_p(d_k) + \sum_{m\neq k} r_p(d_m)}\right] \\
&= 1 - \frac{N_0\alpha}{t(d_k)}\mathbb{E}_{-k}\left[\frac{1}{1+\frac{\sum_{m\neq k} r_p(d_m)}{t(d_k)}}\right] \\
&\approx 1 - \frac{N_0\alpha}{t(d_k)}\mathbb{E}_{-k}\left[e^{-\frac{\sum_{m\neq k} r_p(d_m)}{t(d_k)}}\right] \\
&= 1 - \frac{N_0\alpha}{t(d_k)}\prod_{m\neq k}\mathbb{E}\left[e^{-\frac{r_p(d_m)}{t(d_k)}}\right],
\end{aligned} \tag{5.35}$$

where $t(d_k) = N_0\alpha + r_p(d_k)$ and the approximation is based on $\frac{1}{1+x} \approx e^{-x}$ for $|x| \ll 1$. In Eq. (5.35), with the variational distribution of d_m, i.e., $\rho_m(d)$, we can have

$$\mathbb{E}\left[e^{-\frac{r_p(d_m)}{t(d_k)}}\right] = \sum_{d=0}^{d_{\max}} e^{-\frac{r_p(d)}{t(d_k)}}\rho_m(d). \tag{5.36}$$

Thus, an approximation of the term on the RHS in Eq. (5.22) can be found and the CAVI algorithm can be carried out with the updating rule as

$$\rho_k^*(d_k) \propto e^{N_0\sum_p U_p\left(1-\frac{N_0\alpha}{t(d_k)}\prod_{m\neq k}\mathbb{E}\left[e^{-\frac{r_p(d_m)}{t(d_k)}}\right]\right)+MQ\eta\ln\nu(d_k)}, \tag{5.37}$$

where $U_p = \sum_{m=1}^{M}|u_{m,p}|^2$. Note that the term $e^{MQ\eta\sum_{m\neq k}\mathbb{E}[\ln\nu(d_m)]}$ in $\exp(\mathbb{E}_{-k}[\ln f(d_k\,|\,\mathbf{d}_{-k},\{\mathbf{y}_m\})])$ due to $\Delta(d_1,\dots,d_K)$ is independent of d_k and can be treated as a constant in the updating rule in Eq. (5.37). In particular, with the step size $\mu > 0$, the CAVI algorithm can be implemented using the sequential updating rule as

$$\rho_k^{(i)}(d_k) = \frac{\rho_k^{(i-1)}(d_k) + \mu A_k(i-1)}{\sum_{d=0}^{d_{\max}}\rho_k^{(i-1)}(d) + \mu A_k(i-1)},\ k = 1,\dots,K, \tag{5.38}$$

where

$$A_k(i-1) = e^{N_0\sum_p U_p\left(1-\frac{N_0\alpha}{t(d_k)}\zeta_k(i-1)\right)+MQ\eta\ln\nu(d_k)}. \tag{5.39}$$

Here,

$$\zeta_k(i-1) = \prod_{m<k}\mathbb{E}_{(i)}\left[e^{-\frac{r_p(d_m)}{t}}\right]\prod_{m>k}\mathbb{E}_{(i-1)}\left[e^{-\frac{r_p(d_m)}{t}}\right], \tag{5.40}$$

where $\mathbb{E}_{(i)}$ represents the expectation with the ith variational distribution $\rho_m^{(i)}(d)$.

To check the complexity of the CAVI algorithm, we mainly focus on the number of multiplications. The complexity of the CAVI algorithm in Eq. (5.38) depends on the complexity to find $A_k(i-1)$ and that to obtain $\mathbb{E}_{(i)}\left[e^{-\frac{r_p(d_k)}{t}}\right]$, which is to carry out the expectation in Eq. (5.36) with the updated variational distribution, $\rho_k^{(i)}(d)$. The number of multiplications of the former is proportional to P provided that $\zeta_k(i-1)$ is available, while that of the latter is proportional to $d_{\max}+1$. To update the variational distribution for each k in an iteration, we need $d_{\max+1}$ divisions. Note that since $\{U_p\}$ is obtained prior to iterations, we ignore the complexity to obtain $\{U_p\}$. Consequently, during each iteration, it can be shown that the complexity order is $O(K(P+2(d_{\max}+1))) \approx O(KP)$. Thus, we can see that the complexity of the CAVI algorithm per iteration is proportional to KP.

5.2.3 MCMC Detection Algorithm

According to the ML function, the posterior probability of $\mathbf{d}$ is given by

$$\Pr(\mathbf{d}|\{\mathbf{y}_m\}) \propto f(\mathbf{y}_m \mid \mathbf{d}) \Pr(\mathbf{d}). \tag{5.41}$$

Based on the ML estimation, we use the Markov chain Monte Carlo (MCMC) algorithm to find an approximate solution by obtaining samples of $\mathbf{d} = [d_1, \dots, d_K]^{\mathrm{T}}$ from the target distribution $\Pr(\mathbf{d}|\{\mathbf{y}_m\})$ [76, 77]. We use the Gibbs sampler [78] and the sampling operation can be carried out recursively as [71]

$$\begin{aligned}
d_1^{(n)} &\sim \Pr(d_1|d_2^{(n-1)}, \dots, d_K^{(n-1)}, \{\mathbf{y}_m\}),\\
d_2^{(n)} &\sim \Pr(d_2|d_1^{(n)}, d_3^{(n-1)}, \dots, d_K^{(n-1)}, \{\mathbf{y}_m\}),\\
&\vdots\\
d_k^{(n)} &\sim \Pr(d_k|d_1^{(n)}, \dots, d_{k-1}^{(n)}, d_{k+1}^{(n-1)}, \dots, d_K^{(n-1)}, \{\mathbf{y}_m\}),
\end{aligned} \tag{5.42}$$

where $d^{(n)}$ denotes the nth sample. Letting $\mathbf{d}_{-k}^{(n)} = [d_1^{(n)}, \dots, d_{k-1}^{(n)}, d_{k+1}^{(n-1)}, \dots, d_K^{(n-1)}]^{\mathrm{T}}$, the sampling rule can be formulated as

$$\begin{aligned}
d_k^{(n)} &\sim \Pr(d_k|\mathbf{d}_{-k}^{(n)}, \{\mathbf{y}_m\})\\
&= \frac{f(\{\mathbf{y}_m\}|\mathbf{d}_{-k}^{(n)}, d_k) \Pr(\mathbf{d}_{-k}^{(n)}) \Pr(d_k)}{\sum_d f(\{\mathbf{y}_m\}|\mathbf{d}_{-k}^{(n)}, d_k = d) \Pr(\mathbf{d}_{-k}^{(n)}) \Pr(d_k = d)}\\
&= \frac{f(\{\mathbf{y}_m\}|\mathbf{d}_{-k}^{(n)}, d_k)}{\sum_d f(\{\mathbf{y}_m\}|\mathbf{d}_{-k}^{(n)}, d_k = d)},
\end{aligned} \tag{5.43}$$

where $\mathbf{d}_{-k}^{(n)}$ follows the uniform distribution, i.e., $\Pr(d_k) = \frac{1}{1+d_{\max}}$, $\Pr(\mathbf{d}_{-k}^{(n)}) = (\frac{1}{1+d_{\max}})^{K-1}$.

According to Eq. (5.43), the update of ${d_k}^{(n)}$ can be found as

$$
{d_k}^{(n)} = \underset{d_k \in \{0,\ldots,d_{\max}\}}{\operatorname{argmax}} \frac{{\zeta_k}^{(n)}(d_k)}{\sum_{d=0}^{d_{\max}} {\zeta_k}^{(n)}(d)}, \tag{5.44}
$$

where ${\zeta_k}^{(n)}(d_k) = f(\{\mathbf{y}_m\}|\mathbf{d}_{-k}^{(n)}, d_k)$. After taking the Nth sampling, the samples $\mathbf{d}^{(N)}$ can converge to a peak of the likelihood function. The Markov sequence $\{\mathbf{d}^{(0)}, \ldots, \mathbf{d}^{(N)}\}$ can be obtained with $\Pr(\mathbf{d}^{(n+1)}|\mathbf{d}^{(n)}, \ldots, \mathbf{d}^{(0)}) = \Pr(\mathbf{d}^{(n+1)}|\mathbf{d}^{(n)})$.

However, since the proposed MCMC algorithm may get stuck at a local optimum, the mutation operator in the Genetic Algorithm (GA) is beneficial to protect against this problem by maintaining diversity in the population [79]. Thus, we apply the principle of the mutation operator in GA to the MCMC algorithm to improve the probability to converge to the global optimum. The sampling rule and the sampling process of ${d_k}^{(n)}$ can be regarded as the fitness function and selection process, respectively. Note that the fitness function is used to evaluate the quality of samples in each step, which is

$$
\text{fitness}({d_k}^{(n)}) = \Pr(d_k|\mathbf{d}_{-k}^{(n)}, \{\mathbf{y}_m\}). \tag{5.45}
$$

In the proposed GA-MCMC approach, the value of $\mathbf{d}^{(N)}$ and its fitness value $\mathbf{w}^{(N)}$ are recorded. Then, a mutation operation is carried out to $\mathbf{d}^{(N)}$, which is then updated entirely from the previous ones. After mutation, $\mathbf{d}^{(N)}$ can be used as $\mathbf{d}^{(0)}$ for next N samplings. Finally, $\hat{\mathbf{d}}$ with the largest fitness value of the estimation $\mathbf{d}$ will be selected through G times of iterations. Let $\mathbf{D}$ represent the set of $\mathbf{d}^{(N)}$ and $\mathbf{W}$ represent the fitness value $\mathbf{w}^{(N)}$. The algorithm of GA-MCMC is presented in Algorithm 5.1.

Then, we compare the complexity of the GA-MCMC approach with the ML approach.

In the GA-MCMC algorithm, since the complexity of ${d_k}^{(n)}$ depends on $\sum_{d=0}^{d_{\max}} {\zeta_k}^{(n)}(d)$, considering the number of iterations and mutations, the complexity order becomes $NGK(P - Q + 1)$. It can be observed that the complexity of ML algorithm grows exponentially while the complexity of GA-MCMC grows linearly with K. Thus, if K is large, the complexity saving can be obviously significant.

5.3 Signal Number Estimation and Channel Estimation

In the field of signal processing, the most classical algorithms to estimate the number of targets are information theoretic criteria, including the Akaike Information Criterion (AIC) [80], the Bayesian Information Criterion (BIC) [81], and the Minimum Description Length (MDL) criterion. The MDL criterion is proposed to find the model most likely in the Bayesian sense, which minimizes the

Algorithm 5.1 The GA-MCMC Gibbs Sampler

Initialize the number of samples N, the number of total mutations G, $\mathbf{d}^{(0)}$, $\mathbf{D}$ (the set of $\mathbf{d}^{(N)}$)= $\emptyset$, $\mathbf{W}$ (the set of fitness value $\mathbf{w}^{(N)}$)= $\emptyset$.
Set the number of mutations j=0
while $j < G$ **do**
  for $i = 1$ to N **do**
    Sample $d_1{}^{(i)} \sim \Pr(d_1 | d_2^{(i-1)}, \dots, d_K^{(i-1)}, \{\mathbf{y}_m\})$
    $\dots$
    Sample $d_K{}^{(i)} \sim \Pr(d_K | d_1^{(i)}, \dots, d_{K-1}^{(i)}, \{\mathbf{y}_m\})$
  end for
  Record the sampling results $\mathbf{D} = \mathbf{D} \cup \{\mathbf{d}^{(N)}\}, \mathbf{W} = \mathbf{W} \cup \{\Pr(\mathbf{d}^{(N)} | \{\mathbf{y}_m\})\}$.
  Execute the mutation operations: change $\mathbf{d}^{(N)}$ randomly and set $\mathbf{d}^{(0)} = \mathbf{d}^{(N)}$.
  $j = j + 1$
end while
Select $\hat{\mathbf{d}}$ from $\mathbf{D}$ with the maximum fitness value in $\mathbf{W}$, i.e., $\hat{\mathbf{d}} \leftarrow \max_{\mathbf{W}} \Pr(\mathbf{d}^{(N)} | \{\mathbf{y}_m\})$.
return $\hat{\mathbf{d}}$

total description code-length, comprised of the data likelihood and the model complexity.

To address the problem of finding the number of active signals in RTD estimation, we focus on the determination of the number of active devices K based on the MDL criterion. According to the MDL principle, the model which yields the shortest description length of the data, should be selected from a given data set and a family of probabilistic models.

Suppose that the RTDs are estimated for a given K. Let $\hat{d}_k$ denote the estimate of d_k. Then, from $\{\hat{d}_k\}$, the support of $\mathbf{c}_m$ can be obtained, which is denoted by $\mathcal{I}$. From Eq. (5.6), the non-zero elements of $\mathbf{c}_m$ can be estimated by taking the elements of $\mathbf{z}_m = \boldsymbol{\Psi}^{\dagger}\mathbf{y}_m$ corresponding to the support, i.e., $[\mathbf{z}_m]_{\mathcal{I}}$, or

$$[\hat{\mathbf{c}}_m]_p = \begin{cases} z_{m,p}, & \text{if } p \in \mathcal{I}, \\ 0, & \text{o.w.}, \end{cases} \tag{5.46}$$

where $z_{m,p}$ represents the pth element of $\mathbf{z}_m$. If the inter-arrival time differences or differences of RTDs are greater than or equal to Q, i.e., $|\hat{d}_k - \hat{d}_{k'}| \geq Q$ for $k \neq k'$, from $\hat{\mathbf{c}}_m$, the CIRs of active devices, $\mathbf{v}_{m,k}$, can be obtained as there is no overlapping between the supports of $\mathbf{c}_{m,k}$.

When the inter-arrival time differences are smaller than Q, the CIR of each active device may not be clearly estimated from $\hat{\mathbf{c}}_m$. However, we are still able to determine the number of active devices, K, since the $\hat{\mathbf{c}}_m$'s are required to form an objective function to determine K. With $\{\hat{d}_k\}$, and $\{\hat{\mathbf{c}}_m\}$ (for given K), denote by

$f(\{\mathbf{y}_m\} \mid \{\hat{d}_k\}, \{\hat{\mathbf{c}}_m\}, K)$ the generalized likelihood function of K. Then, based on the MDL criterion [82], the problem to determine K can be formulated as

$$K^* = \underset{K \in \{0,\dots,K_{\max}\}}{\operatorname{argmin}} \; -\ln f(\{\mathbf{y}_m\} \mid \{\hat{d}_k\}, \{\hat{\mathbf{c}}_m\}, K) + \frac{N_{\text{par}}(K)}{2} \ln(N_{\text{data}}), \tag{5.47}$$

where $N_{\text{par}}(K)$ and N_{data} represent the number of parameters for given K and the number of data samples, respectively, and $K_{\max}$ is the maximum number of active devices. For given K, since there are K RTDs and $2MQK$ channel coefficients (for $\{\mathbf{c}_m\}$, where each element is complex-valued), we have

$$N_{\text{par}}(K) = K(1 + 2MQ).$$

In addition, since the elements of $\mathbf{y}_m$ are complex-valued, N_{data} becomes $2LM$. The second term on the RHS in Eq. (5.47) is the penalty function and increases with K, which is seen as the model order, whereas the first term on the RHS in Eq. (5.47), i.e., the negative likelihood function, is generally a non-increasing function of K.

5.4 Simulation Results

5.4.1 CAVI Simulation Results

We present simulation results with the distance function as

$$\nu(d)^{\eta} = \left(\frac{d+1}{d_{\max}+1} \right)^{-\eta}, \; d \in \{0, 1, \dots, d_{\max}\}. \tag{5.48}$$

The gain of this distance function becomes unity when $d = d_{\max}$ for normalization and the path loss exponent η is set to 3. For the PDP, we consider the exponential decay function as

$$\overline{\sigma}_q^2 = \exp(-\beta q), \; q = 0, \dots, Q-1,$$

where the exponent β is set to 0.8. It is also assumed that s_l is an independent binary phase shift keying (BPSK) symbol, i.e., $s_l \in \{-1, +1\}$, and the signal-to-noise ratio (SNR) is defined as

$$\text{SNR} = \frac{||\mathbf{s}||^2}{N_0} = \frac{1}{\alpha N_0}. \tag{5.49}$$

Thus, from Eq. (5.48), this SNR is normalized for the device of the maximum RTD, i.e., $d_{\max}$. We consider a random initialization for the CAVI algorithm where $\rho_k(d)^{(0)}$ is randomly chosen from $\mathcal{P}$.

Prior to presenting the main simulation results, we compare the performance of the CAVI algorithm with that of the conventional method in [74]. We consider the case of $K = 2$ with $L = 64$, $P = 16$, $Q = 3$, and SNR $= 10$ dB. Furthermore, we assume that the RTDs are fixed as $d_1 = 5$ and $d_2 = 8$ so that there is no CIR overlapping. For the CAVI algorithm, we have $(\mu, N_{\text{iter}}) = (1, 20)$. For the correlator-based detector [74], we choose the largest elements of $\sum_{m=1}^{M} |z_{m,p}|^2$ to estimate RTDs. In particular, we obtain the estimates of RTDs of the correlator-based detector as

$$\hat{d}_1 = \underset{p\in\{0,\ldots,7\}}{\text{argmax}} \sum_{m=1}^{M} |z_{m,p}|^2, \ \hat{d}_2 = \underset{p\in\{8,\ldots,15\}}{\text{argmax}} \sum_{m=1}^{M} |z_{m,p}|^2. \tag{5.50}$$

The normalized frequencies[2] of $\hat{d}_k$ are presented in Fig. 5.2, where we can clearly see that the performance of the CAVI algorithm outperforms that of the conventional correlator-based detector. It is also shown that the CAVI algorithm can successfully estimate the correct RTDs $(d_1 = 5, d_2 = 8)$, with a high probability, i.e., 0.98.

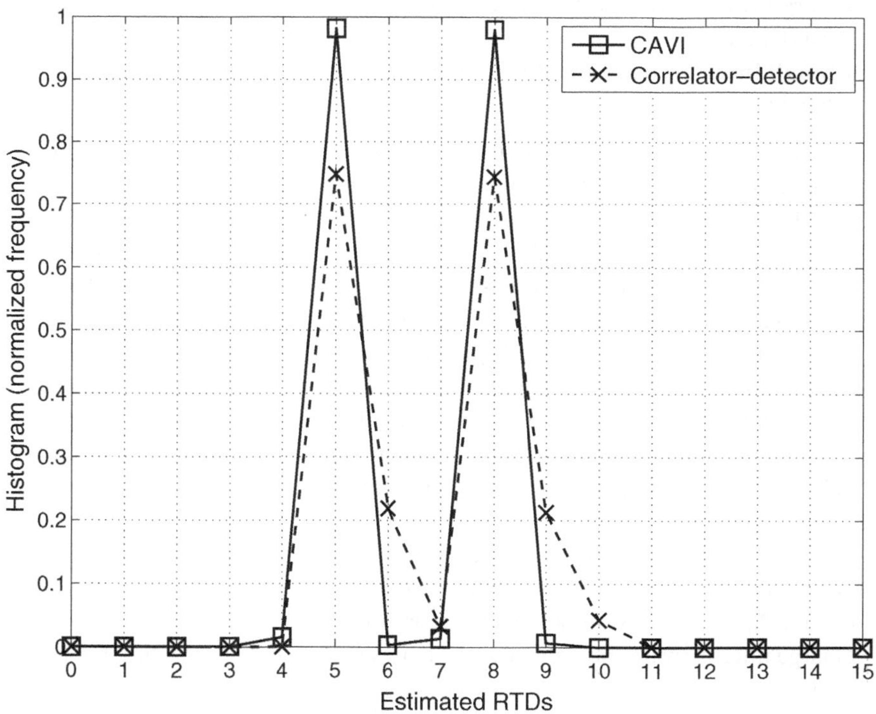

Figure 5.2 Normalized frequencies of $\hat{d}_k$ of the CAVI algorithm and the correlator-based detector when $K = 2$ $(d_1 = 5, d_2 = 8)$, $L = 64$, $P = 16$, $Q = 3$, $M = 2$, SNR = 10 dB, and $(\mu, N_{\text{iter}}) = (1, 20)$.

2 The normalized frequency of $\hat{d}_k$ is $\frac{\text{Number of occurrences of } \hat{d}_k}{\text{Number of trials}}$.

We now present the main simulation results with random RTDs to evaluate the performance of the CAVI algorithm under various conditions. The RTD, d_k, is assumed to be uniformly and independently chosen from $\{0, \dots, d_{\max}\}$ at random, i.e., $\Pr(d_k) = \frac{1}{1+d_{\max}} = \frac{1}{P-Q+1}, k = 1, \dots, K$. In addition, in order to analyze the RTD estimation performance under the complex condition of different number of active signals, we consider the mean of the number of successfully recovered RTDs MNSRD as a performance measure for the RTD estimation, which is given by

$$\text{MNSRD} = \mathbb{E}[|\{d_1, \dots, d_K\} \backslash \{\hat{d}_1, \dots, \hat{d}_K\}|] (\leq K),$$

where $\backslash$ represents the set difference ($A \backslash B = \{x \mid x \in A \, \text{and} \, x \notin B\}$). Clearly, if MNSRD $\ll 1$, we can see that the RTDs are correctly estimated with a high probability. In order to take into account the impact of the inter-arrival time difference on the MNSRD for the case of $K = 2$, we consider two cases:

- **Large inter-arrival time difference**: The case of $|d_1 - d_2| \geq Q$, i.e., no CIR overlapping;
- **Small inter-arrival time difference**: The case of $|d_1 - d_2| < Q$, i.e., CIR overlapping.

Note that in the case of small inter-arrival time difference or CIR overlapping, the estimation error of RTD can be high due to the interference.

In Fig. 5.3, with $K = 2$, the performance of the CAVI algorithm for different numbers of iterations, denoted by N_{iter}, is shown when $L = 64$ and $P = 16$, $Q = 3$, $M = 2$, SNR $= 10$ dB, and $\mu = 1$. It is shown that the (conditional) MNSRD of CAVI for the case of large inter-arrival time difference or no CIR overlapping decreases with the number of iterations and approaches that of ML, i.e., the CAVI algorithm can provide a near ML performance if $|d_1 - d_2| \geq Q$ with a sufficiently large number of iterations (e.g., $N_{\text{iter}} \geq 15$). On the other hand, the (conditional) MNSRD of CAVI for the case of CIR overlapping (after convergence) is slightly larger than that of ML. This shows that the CAVI algorithm cannot provide a near ML performance if $|d_1 - d_2| < Q$.

Figure 5.4 shows the performance of the CAVI algorithm to estimate RTDs for different values of step size μ, when $K = 2, L = 64, P = 16, Q = 3, M = 2$, SNR $= 10$ dB, and $N_{\text{iter}} = 20$. It is shown that the (conditional) MNSRD of CAVI for the case of large inter-arrival time difference decreases with μ and approaches that of ML.

From Figs. 5.3 and 5.4, we can see that the CAVI algorithm can provide a reasonably good performance with $N_{\text{iter}} \geq 15$ and $\mu \geq 0.6$, and its performance can be close to that of ML for the case of no CIR overlapping, i.e., $|d_1 - d_2| \geq Q$. Furthermore, even if $|d_1 - d_2| < Q$, the CAVI algorithm has a slightly worse performance than ML.

The impact of SNR on the performance of the CAVI algorithm and the ML approach is shown in Fig. 5.5 when $K = 2$, $L = 64$, $P = 16$, $Q = 3$, $M = 2$, and

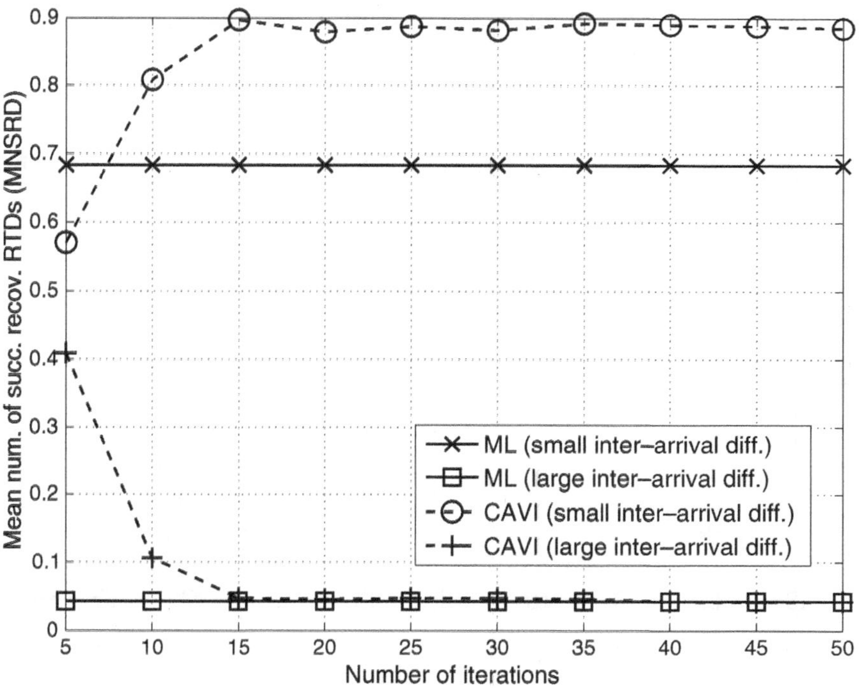

Figure 5.3 Performance of the CAVI algorithm to estimate RTDs for different numbers of iterations when $K = 2, L = 64, P = 16, Q = 3, M = 2$, SNR = 10 dB, and $\mu = 1$.

$(\mu, N_{\text{iter}}) = (1, 20)$. In general, the MNSRD decreases with SNR, and the conditional MNSRD for the case of large inter-arrival time difference approaches 0 in both the ML approach and the CAVI algorithm. However, for small inter-arrival time difference, i.e., $|d_1 - d_2| < Q$, the MNSRD does not approach 0 although the SNR is high (e.g., 30 dB) due to the interference from the other active device, although the background noise is negligible.

Figure 5.6 shows the performance of the CAVI algorithm and the ML approach for various numbers of paths, Q, when $K = 2, L = 64, P = 16$, SNR = 10 dB, $M = 2$, and $(\mu, N_{\text{iter}}) = (1, 20)$. It is shown that the MNSRD increases with Q. Note that since $P = 16$, if $Q = 8$, CIR overlapping cannot be avoided. Thus, when $Q = 8$, the conditional MNSRD for the case of large inter-arrival time difference cannot be obtained, which is however shown to be 0 in Fig. 5.6 for convenience. Furthermore, since the probability that two CIRs are overlapped, increases with Q, the average MNSRD becomes the conditional MNSRD for the case of small inter-arrival time difference as Q increases. Clearly, we can see that a large Q is not desirable for a reliable estimate of RTD.

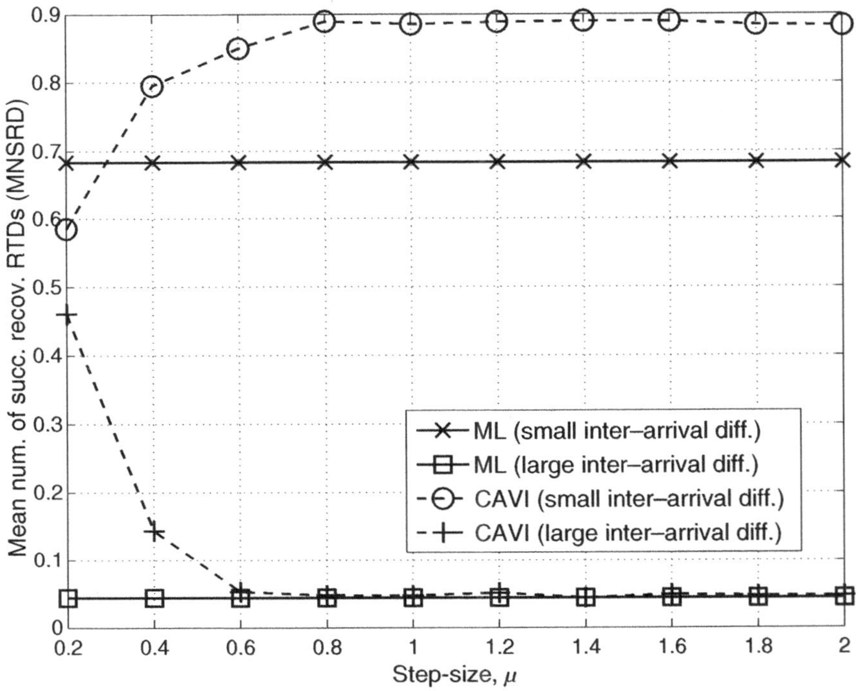

Figure 5.4 Performance of the CAVI algorithm to estimate RTDs for different values of step size μ, when $K = 2$, $L = 64$, $P = 16$, $Q = 3$, $M = 2$, SNR = 10 dB, and $N_{\text{iter}} = 20$.

The impact of the number of antennas, M, on the performance, is shown in Fig. 5.7 when $K = 2, L = 64, P = 16, \text{SNR} = 10\,\text{dB}, Q = 3$, and $(\mu, N_{\text{iter}}) = (1, 20)$. In Fig. 5.7, we can see that the conditional MNSRD for the case of large inter-arrival time difference can approach 0 as M increases, whereas the conditional MNSRD for the case of small inter-arrival time difference is almost independent of M. Since the interference due to CIR overlapping exists regardless of the number of antennas, the increase of M does not significantly help reduce the MNSRD in the case of small inter-arrival time difference.

From Figs. 5.5–5.7, we can see that the performance can vary considerably depending on CIR overlapping. In general, the RTD estimation is quite reliable if there is no CIR overlapping, and the channel estimation can be successfully carried out as the CIR of each active device can be recovered without any interference. In Fig. 5.8, the conditional normalized mean-squared error (MSE) of the estimated composite CIR in Eq. (5.46), defined as

$$\text{Normalized MSE} = \frac{\mathbb{E}[||\mathbf{c}_m - \hat{\mathbf{c}}_m||^2]}{\mathbb{E}[||\mathbf{c}_m||^2]},$$

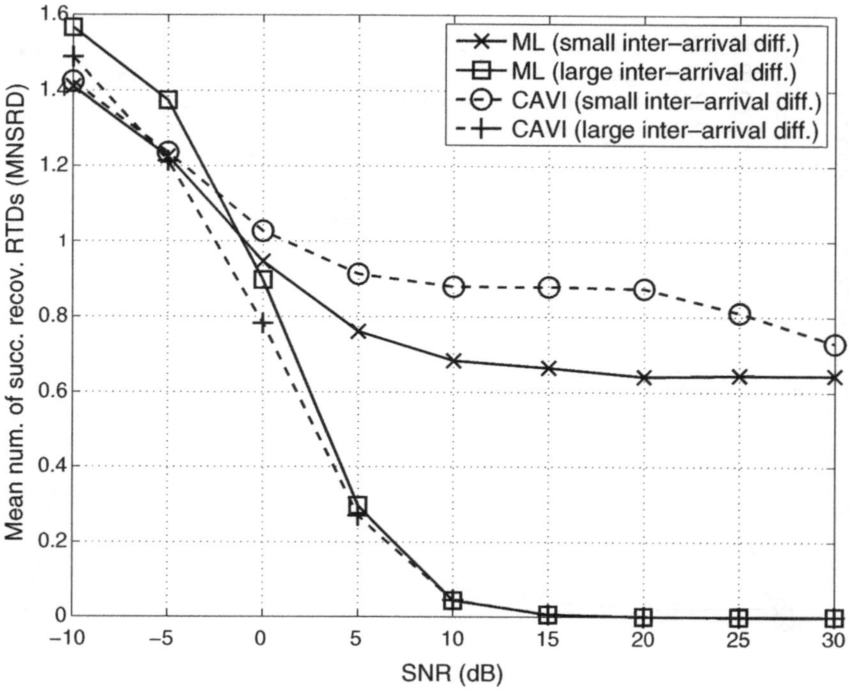

Figure 5.5 Performance of the CAVI algorithm and the ML approach for different values of SNR when $K = 2$, $L = 64$, $P = 16$, $Q = 3$, $M = 2$, and $(\mu, N_{\text{iter}}) = (1, 20)$.

is shown for the case of no CIR overlapping when $K = 2$, $L = 64$, $P = 16$, and $(\mu, N_{\text{iter}}) = (1, 20)$. The normalized MSEs of the channel estimate as functions of the number of multipaths, Q, SNR, and the number of antennas, M. In general, we can see that the channel estimation can be reliably performed as the normalized MSE is close to 10^{-2} with moderate SNRs (e.g., greater than or equal to 10 dB). As expected, the normalized MSE of channel estimate decreases with SNR, Q, and M.

As mentioned earlier, we have considered the RTD estimation to determine the number of active devices or signals, K. Based on the MDL criterion in Eq. (5.47), simulations are carried out to find the conditional probability of correct estimation of K for given $K \in \{0, \dots, 3\}$, i.e., $K_{\max} = 3$, for different values of SNR when $L = 64$, $P = 16$, $Q = 3$, and $M = 2$. The CAVI algorithm is used with $(\mu, N_{\text{iter}}) = (1, 20)$ to obtain the estimates of RTDs with 4 possible values of K, i.e., $K \in \{0, \dots, 3\}$.

From the simulation results presented in Fig. 5.9, we can see that K can be reliably estimated when $K = 0$, i.e., there is no active device, as a reference, for a wide range of SNR. However, as K increases, the conditional probability of correct

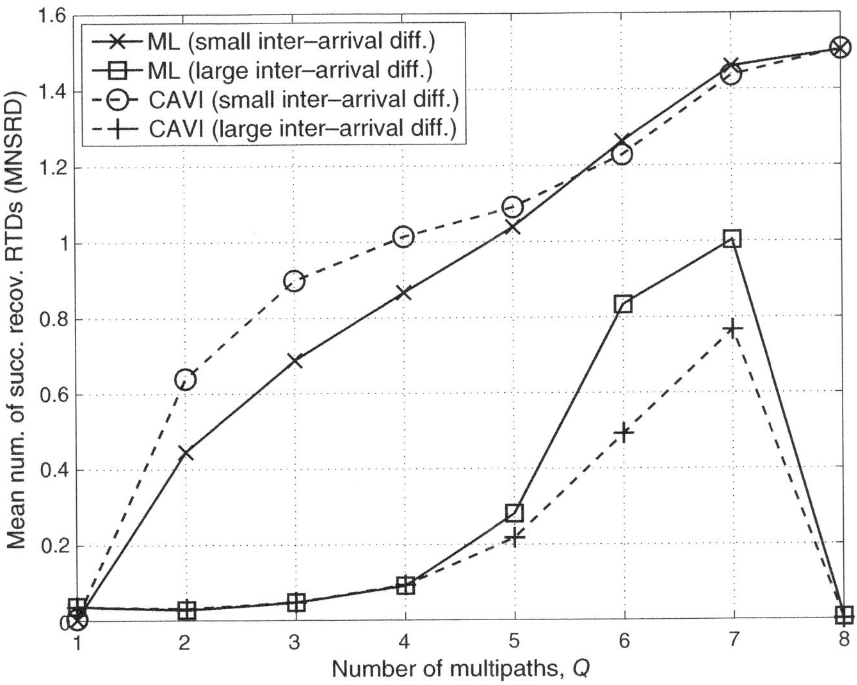

Figure 5.6 Performance of the CAVI algorithm and the ML approach for different numbers of multipaths, Q, when $K = 2$, $L = 64$, $P = 16$, SNR = 10 dB, $M = 2$, and $(\mu, N_{\text{iter}}) = (1, 20)$.

estimation of K decreases, because the event of CIR overlapping becomes more frequent as K increases. In addition, we observe that for a larger K, a higher SNR is required to successfully estimate K with a high probability. It is noteworthy that in practice, since the number of active devices per RB might be small with a high probability,[3] it is desirable to have a high (conditional) probability of correct estimation of K when K is small [67]. From this perspective, we can see that the results in Fig. 5.9 can be reasonable because the *average* probability of an accurate estimate of K can be high.

Figure 5.10 (a) shows the conditional probabilities of correct estimation of K for given $K \in \{0, \ldots, 3\}$ as functions of the number of multipaths, Q, when $L = 64$, $P = 16$, SNR = 10 dB, $M = 2$, and $(\mu, N_{\text{iter}}) = (1, 20)$. While the conditional probability of correct estimation of K decreases with K as mentioned earlier, we can observe that the conditional probability of correct estimation of K

3 The maximum probability of $K \geq 3$ is 0.056, when the total number of arrived devices is 50 000.

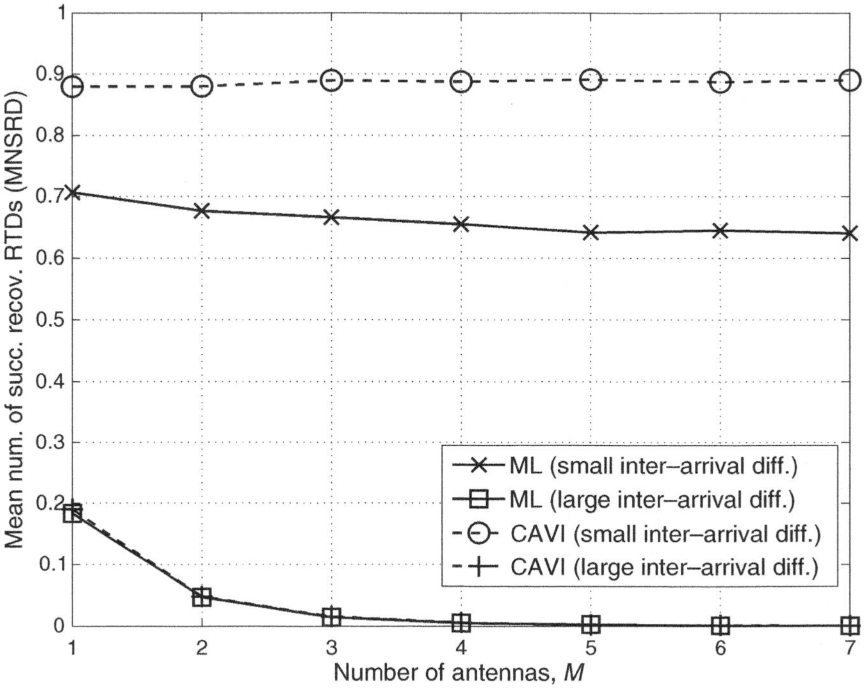

Figure 5.7 Performance of the CAVI algorithm and the ML approach for different numbers of antennas, M, when $K = 2$, $L = 64$, $P = 16$, SNR = 10 dB, $Q = 3$, and $(\mu, N_{\text{iter}}) = (1, 20)$.

decreases with Q. Since the probability of the event of CIR overlapping increases with Q for a fixed P, it is more difficult to correctly decide K RTDs, which leads to the increase in the conditional probability of correct estimation of K with Q. To overcome this problem, we can increase the length of CP, which can effectively decrease the probability of the event of CIR overlapping for a fixed Q, at the cost of more bandwidth, i.e., with more subcarriers. We can confirm the performance improvement by increasing P in Fig. 5.10 (b).

5.4.2 MCMC Simulation Results

Since the GA-MCMC can provide the optimal solution to the ML approach asymptotically, we analyze the performance of the GA-MCMC method. Suppose that s_l is an independent BPSK symbol, i.e., $s_l \in \{-1, +1\}$. Let $\mathbf{u}_m = \mathbf{\Psi}^{\text{H}}\mathbf{y}_m = \mathbf{F}_P^{\text{H}}\mathbf{S}^{\text{H}}\mathbf{y}_m$ and we define

$$r_p(d) = \begin{cases} v(d)^{-\eta}\overline{\sigma}_{p-d}^{2}, & \text{if } p \in \{d, \dots, d+Q-1\}, \\ 0, & \text{o.w.} \end{cases} \tag{5.51}$$

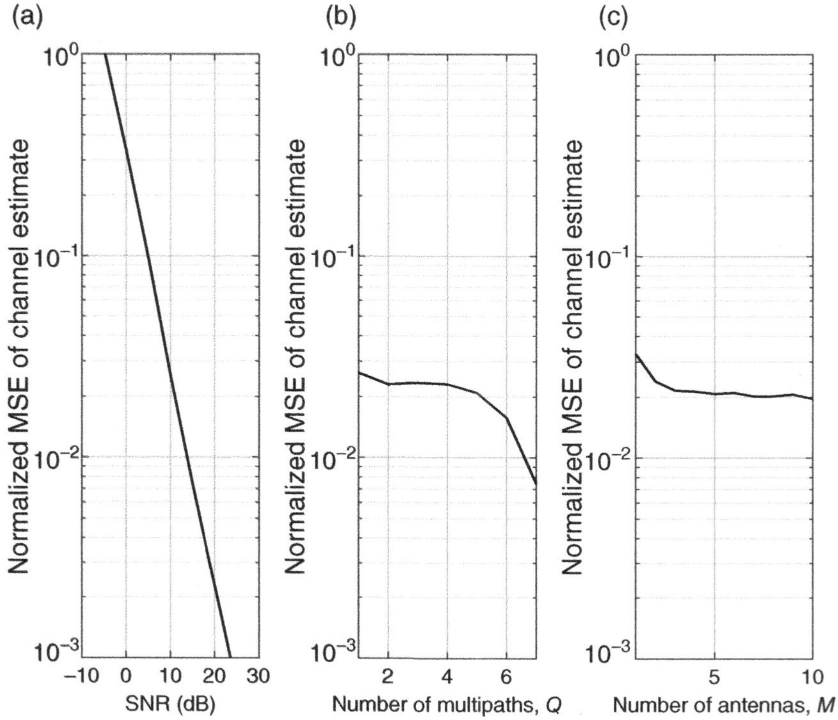

Figure 5.8 Conditional normalized MSE of the estimated composite CIR (without CIR overlapping) when $K = 2$, $L = 64$, $P = 16$, and $(\mu, N_{\text{iter}}) = (1, 20)$: (a) as a function of SNR with $Q = 3$ and $M = 2$; (b) as a function of Q with SNR = 10 dB and $M = 2$; (c) as a function of M with SNR = 10 dB and $Q = 3$.

In addition, in order to analyze the RTD estimation performance, we consider the probability of the number of correct recovered RTDs (P_c), which is given by

$$P_c = \frac{|\{d_1, \ldots, d_K\} \cap \{\hat{d}_1, \ldots, \hat{d}_K\}|}{|\{d_1, \ldots, d_K\}|}, \tag{5.52}$$

where $\hat{d}$ denotes the estimate of d.

There are two cases for the estimation of $\mathbf{d}$:

- the completely correct estimation of $\mathbf{d}$ ($\hat{\mathbf{d}} = \mathbf{d}$),
- the partially correct estimation of $\mathbf{d}$ ($\hat{\mathbf{d}} \neq \mathbf{d}$ and $\hat{\mathbf{d}} \cap \mathbf{d} \neq \emptyset$).

For the first case, let $\mathbf{d}_r$ denote a random vector in $\mathcal{D}$, where $\mathcal{D} = \{\mathbf{d} = [d_1, \ldots, d_K]^{\text{T}} | d_k \in \{0, 1, \ldots, d_{\max}\}\}$. For convenience, we use the logarithm of the

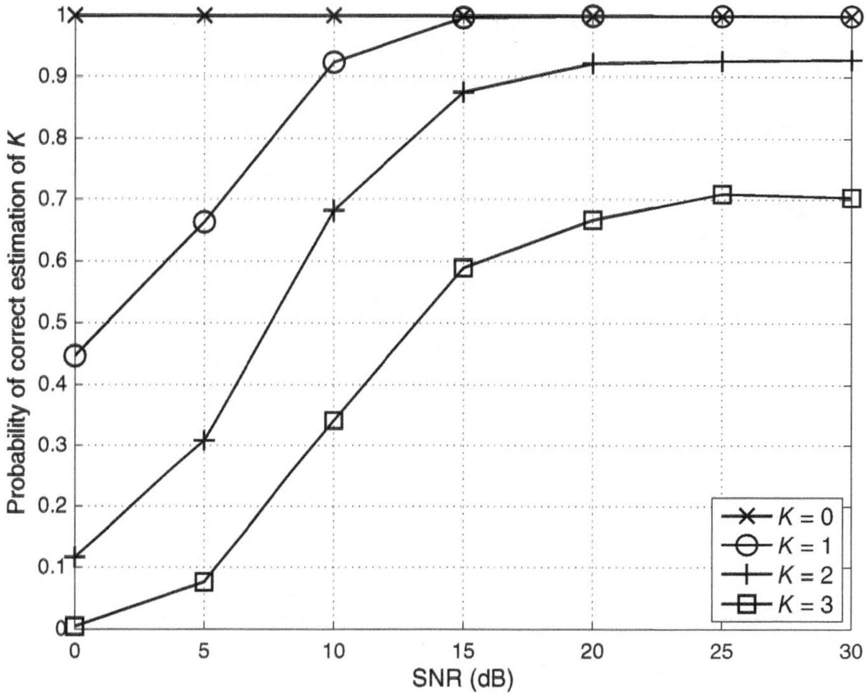

Figure 5.9 Conditional probabilities of correct estimation of K for given $K \in \{0, \ldots, 3\}$ as functions of SNR when $L = 64$, $P = 16$, $Q = 3$, $M = 2$, and $(\mu, N_{\text{iter}}) = (1, 20)$.

likelihood function, while the correct estimation probability is shown as

$$\begin{aligned} \text{P}_{\text{ct}} &= \sum_{\mathbf{d}_r \in \mathcal{D}} \Pr(\ln f(\{\mathbf{y}_m\} \mid \hat{\mathbf{d}}) > \ln f(\{\mathbf{y}_m\} \mid \mathbf{d}_r)) \Pr(\mathbf{d}_r) \\ &= \sum_{\mathbf{d}_r \in \mathcal{D}} \Pr\left(N_0 \sum_p |u_p|^2 \frac{\alpha N_0 \sum_k (r_p(\hat{d}_k) - r_p(d_{rk}))}{C_d} \right. \\ &\qquad \left. > MQ\eta \sum_k \ln \frac{\nu(d_{rk})}{\nu(\hat{d}_k)} \right) \Pr(\mathbf{d}_r), \end{aligned} \tag{5.53}$$

where $C_d = (\alpha N_0 + \sum_k r_p(\hat{d}_k))(\alpha N_0 + \sum_k r_p(d_{rk}))$, $\alpha = \frac{1}{|s_l|^2 L} = \frac{1}{\|\mathbf{s}\|^2}$, and $\hat{d}_k$, d_{rk} represent the kth element of $\hat{\mathbf{d}}$, $\mathbf{d}_r$, respectively. For the second case, let $\mathcal{D}_r$ denote the set of $\mathbf{d}_r$with $\mathbf{d}_r \cap \mathbf{d} \neq \emptyset$ and *num* denote the number of elements in the

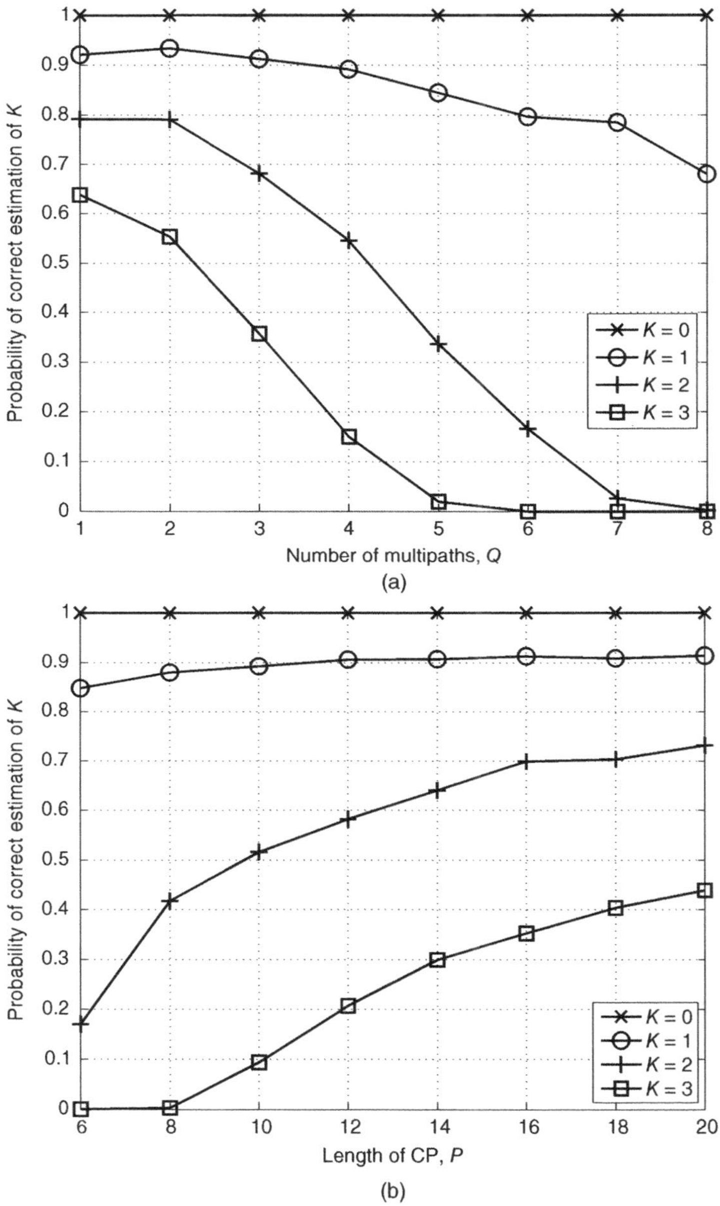

Figure 5.10 Conditional probabilities of correct estimation of K for given $K \in \{0, \ldots, 3\}$ when $L = 64$, SNR = 10 dB, $M = 2$, and $(\mu, N_{\text{iter}}) = (1, 20)$: (a) as functions of the number of multipaths, Q, when $P = 16$; (b) as functions of the length of CP, P, when $Q = 3$.

intersection ($\mathbf{d}_r \cap \mathbf{d}$), then we have

$$
\begin{aligned}
\mathrm{P}_{\mathrm{num}} &= 1 - \sum_{\hat{\mathbf{d}} \in D_r} \Pr\left(\ln f\left(\{\mathbf{y}_m\} \middle| \mathbf{d}\right) > \ln f\left(\{\mathbf{y}_m\} \middle| \hat{\mathbf{d}}\right)\right) \Pr\left(\hat{\mathbf{d}}\right) \\
&\quad - \sum_{\hat{\mathbf{d}} \notin D_r} \Pr\left(\ln f\left(\{\mathbf{y}_m\} \middle| \mathbf{d}\right) < \ln f\left(\{\mathbf{y}_m\} \middle| \hat{\mathbf{d}}\right)\right) \Pr\left(\hat{\mathbf{d}}\right) \\
&= 1 - \sum_{\hat{\mathbf{d}} \in D_r} \Pr\left(N_0 \sum_p |u_p|^2 \frac{\alpha N_0 \sum_k \left(r_p(d_k) - r_p\left(\hat{d}_k\right)\right)}{C_d} > MQ\eta \sum_k \ln \frac{\nu\left(\hat{d}_k\right)}{\nu(d_k)} \right) \Pr\left(\hat{\mathbf{d}}\right) \\
&\quad - \sum_{\hat{\mathbf{d}} \notin D_r} \Pr\left(N_0 \sum_p |u_p|^2 \frac{\alpha N_0 \sum_k \left(r_p(d_k) - r_p\left(\hat{d}_k\right)\right)}{C_d} < MQ\eta \sum_k \ln \frac{\nu\left(\hat{d}_k\right)}{\nu(d_k)} \right) \Pr\left(\hat{\mathbf{d}}\right).
\end{aligned}
\tag{5.54}
$$

Thus, P_c can be simplified as

$$
P_c = \mathrm{P}_{\mathrm{ct}} + \sum_{\mathrm{num}=1}^{K-1} \mathrm{P}_{\mathrm{num}} \lambda_{\mathrm{num}}, \tag{5.55}
$$

where λ_{num} is the weight parameter of $\mathrm{P}_{\mathrm{num}}$ and $\lambda_{\mathrm{num}} = \frac{\mathrm{num}}{K}$. For example, while $K = 2$, we have $P_c = \mathrm{P}_{\mathrm{ct}} + \frac{1}{2}\mathrm{P}_1$.

Assuming that d_k is randomly chosen from $\{0, \ldots, d_{\max}\}$, we present the simulation results with respect to P_c to illustrate the performance of the proposed method. For comparisons, the following methods are considered:

- **MCMC**: RTD estimation using the MCMC approach with Gibbs Sampler;
- **GA-MCMC**: RTD estimation using the GA-based MCMC approach with Gibbs Sampler;
- **ML**: RTD estimation using the ML approach with Exhaustive Search.

Figure 5.11 shows P_c of the three methods with iterations. It is shown that P_c of GA-MCMC for the case of large inter-arrival time difference first increases dramatically and then becomes saturated as the number of samples increases. It also indicates that GA-MCMC has a higher P_c than MCMC, which provides the near ML performance.

The impact of the number of mutations G on P_c is shown in Fig. 5.12. It can be observed that P_c of GA-MCMC for the case of large inter-arrival time difference approaches that of ML as G increases. It shows that by carrying out the mutation operations, GA-MCMC is able to find a global optimal solution with high probability.

Figure 5.13 shows the comparison of P_c obtained from analytical results and numerical results. It is shown that the analytical results match well with the numerical ones. It can be shown that P_c increases with SNR in general. In particular, P_c approaches 1 in GA-MCMC and ML for large inter-arrival time difference.

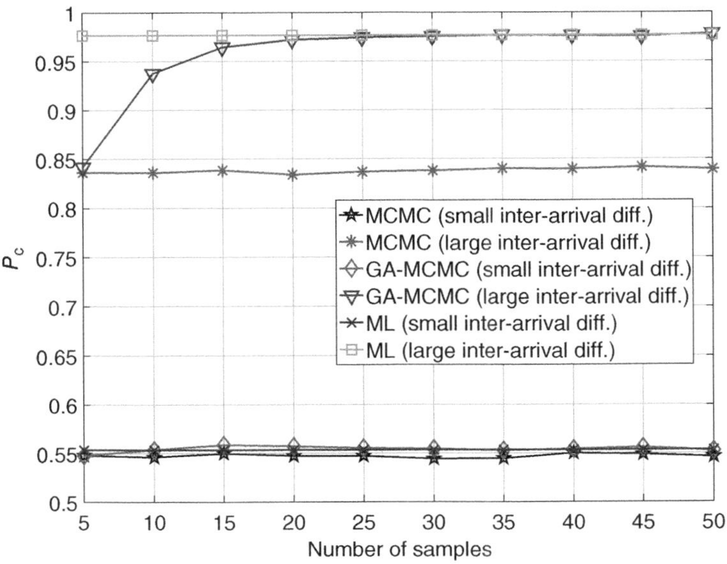

Figure 5.11 P_c with numbers of samples under $L = 64$, $P = 16$, $Q = 3$, $M = 2$, and SNR = 10 dB.

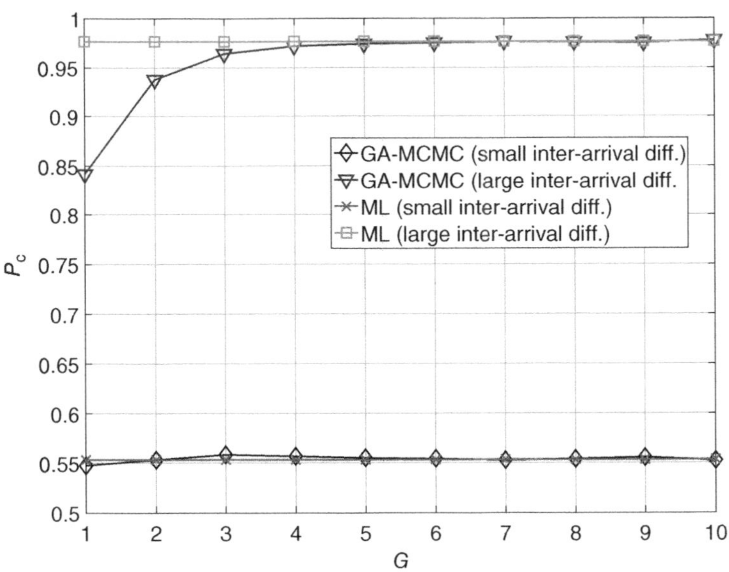

Figure 5.12 P_c with numbers of mutations under $L = 64$, $P = 16$, $Q = 3$, $M = 2$, and SNR = 10 dB.

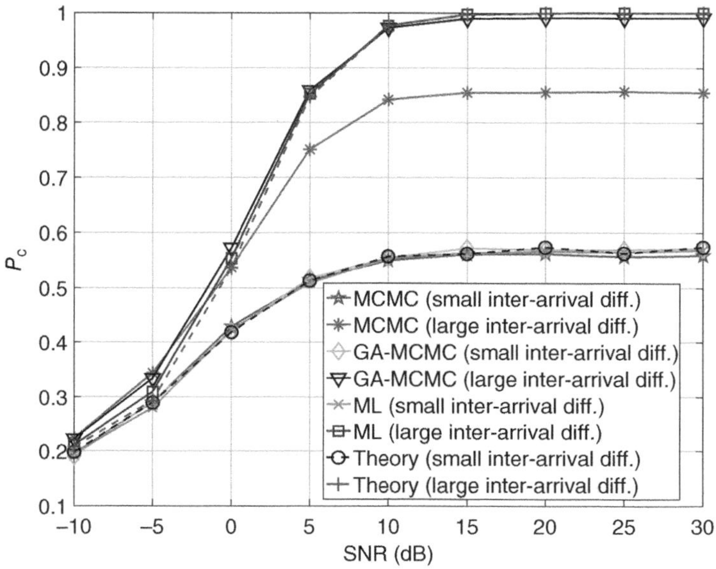

Figure 5.13 P_c with SNR under $K = 2$, $L = 64$, $P = 16$, $Q = 3$, and $M = 2$.

Therefore, the GA-MCMC method can improve the performance, especially under the condition of large inter-arrival time difference, i.e., no CIR overlapping.

5.5 Conclusion and Remarks

In this chapter, we introduced signal estimation methods to provide the estimates of the arrival time of multiple signals in NORA. We first considered the ML approach to estimate RTDs, and then derived low-complexity approaches based on VI and GA-MCMC, respectively. With the estimated RTDs, the determination of the number of active devices or signals has been studied with the MDL criterion.

While we only focused on the determination of the number of multiple signals, K, through the estimation of RTDs, for collision detection, there are other issues to be studied in the future. For example, the impact of the error of RTD estimates on the performance of NORA can be investigated. To mitigate CIR overlapping, access control that assigns devices to RBs depending on their RTDs might be an interesting topic.

After discussing Internet of Things (IoT) multiple access detection with signal collision, we will introduce some special cases in Chapter 6, e.g., backscatter communications, where IoT devices transmit their data through harvesting and reflecting the energy of radio frequency sources, which may cause the signal superposition.

6

Detection and Division for Backscatter Signals

- In machine-type communication (MTC), there might be power-limited or battery-free devices (namely parasite devices) such as radio frequency identification (RFID) tags, which can use backscatter communications to transmit their data through harvesting and reflecting the energy of radio frequency sources (namely host devices). Two types of backscatter communication techniques are most widely applied in MTC systems, which are the bistatic backscatter communication system (BBCS) and the ambient backscatter communication system (ABCS). Since radio frequency sources of BBCS are dedicated carrier emitters, e.g., power beacons, the cost can be relatively high. While ABCS uses the already-available radio frequency (RF) sources for backscatter communications, it can reduce costs and improve spectrum efficiency as no dedicated energy supply is deployed in the system. However, the unpredictability and dynamics of the ambient signal make it a direct interference to the receiver, enabling the detection of signals to play a pivotal role in the ABCS. In this chapter, we introduce an approach to detect the presence of parasite devices and separate parasite signals from host signals [83].

6.1 System Model

In most cases, frequency ranges for backscatter communications with passive RFID tags are not high (e.g., high frequency (HF) of 13.56 MHz or ultrahigh frequency (UHF) of 856–960.56 MHz) [84], while there is also microwave RFID operating at 2.45 GHz [85]. Thus, to support backscatter communications for battery-free devices such as RFID tags within MTC [86, 87], it would be desirable for 5G systems to operate at a sub-gigahertz band. In fact, in long-term evolution (LTE), Narrowband Internet of Things (NB-IoT) operates at

IoT Signal Detection, First Edition. Rui Han, Jingjing Wang, Lin Bai, and Jianwei Liu.

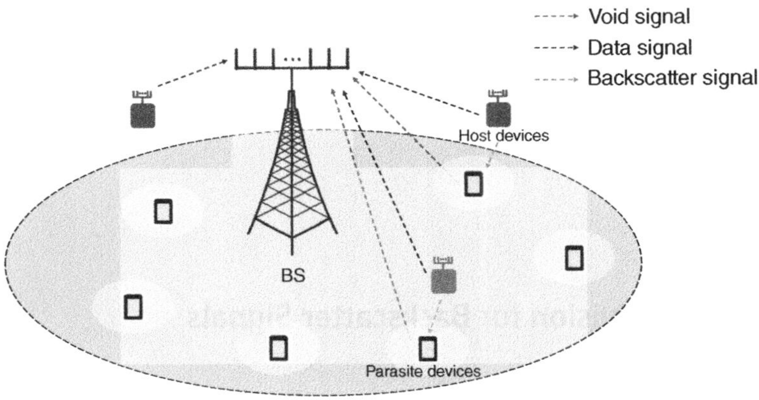

Figure 6.1 Backscatter communication system.

sub-gigahertz frequency ranges, i.e., 700, 800, and 900 MHz. It is also noteworthy that long-range backscatter communications become possible using multiple antennas or multiple-input multiple-output (MIMO) [88–92].

Firstly, we describe the structure of backscatter communication systems. Suppose that a backscatter communication system consists of a base stations (BS) and multiple devices. It is assumed that the BS is equipped with M antenna elements, where M is usually large for massive MIMO [18]. As illustrated in Fig. 6.1, there are two different types of devices: host devices and parasite devices. Considering the security of information, host devices help multiple parasite devices transmit signals to the BS directly [93, 94].

In order to minimize signaling overhead and access delay, the grant-free random access (RA) scheme is used in backscatter communication systems, which skips the resource scheduling of BS and enables the simultaneous transmissions of multiple signals [95]. The process is presented as follows. Firstly, the BS broadcasts the system information, from which each host device or each parasite device estimates its channel gain, and then employs fractional power control (FPC) to compensate for the path loss, where FPC is an open-loop power control technology used in the LTE cellular networks. It is noteworthy that to facilitate the subsequent analysis, the path loss compensation factor of FPC is set to 1. Once a host device or a parasite device attempts to access the network, it randomly picks a preamble from a predefined preamble pool assigned to the corresponding cell. Followed by the data, the preamble is transmitted to the BS through the physical random access channel (PRACH) [55].

In grant-free RA, there are two phases. The first one is the phase of preamble transmission and the second one is the phase of data transmission. In the following, we present a transmission model with host devices and parasite devices for grant-free RA in detail. The first one is the phase of preamble transmission and the second one is the phase of data transmission.

1) **Grant-free RA with host devices:** In the phase of preamble transmission, a host device with data to send firstly becomes active, and then selects a preamble from the preamble pool randomly, which is shown as

$$C = \{\mathbf{c}_1, \dots, \mathbf{c}_L\}. \tag{6.1}$$

Here, $\mathbf{c}_l \in \mathbb{C}^{\tilde{L}\times 1}$ denotes the lth preamble, which is an orthonormal sequence of length $\tilde{L}$. Thus, we assume that $L = \tilde{L}$. After sending a preamble, the host device sends its data packet in the phase of data transmission. For convenience, we assume that devices are synchronized and data packets of host devices have the same length.

In the phase of preamble transmission, the signal received at the BS can be given by

$$\mathbf{Y} = \sum_{k=1}^{K} \mathbf{h}_k \sqrt{P_k} \mathbf{c}_{l(k)}^{\mathrm{T}} + \mathbf{N} \in \mathbb{C}^{M\times\tilde{L}}, \tag{6.2}$$

where K denotes the number of active host devices, $\mathbf{h}_k \in \mathbb{C}^{M\times 1}$ denotes the channel vector from the kth host device to the BS, $l(k)$ represents the index of the preamble chosen by the kth host device, P_k represents the transmit power of the kth host device, and $[\mathbf{N}]_{m,\tilde{l}} \sim \mathcal{CN}(0, N_0)$ is the background noise at the mth antenna and the $\tilde{l}$th preamble symbol duration.

In the phase of data transmission, each active host device transmits its data packet of length D to the BS. The signal received at the BS becomes

$$\mathbf{Z} = \sum_{k=1}^{K} \mathbf{h}_k \sqrt{P_k} \mathbf{a}_k^{\mathrm{T}} + \tilde{\mathbf{N}} \in \mathbb{C}^{M\times D}, \tag{6.3}$$

where $\mathbf{a}_k$ denotes the data packet from the kth host device and $[\tilde{\mathbf{N}}]_{m,d} \sim \mathcal{CN}(0, N_0)$ denotes the background noise at the mth antenna in the dth data symbol duration. Throughout the paper, we consider the following assumptions.

A1) As in [96], P_k is decided to be inversely proportional to the distance between the BS and the kth active host device via power control so that

$$\mathbf{h}_k \sqrt{P_k} = \mathbf{v}_k \sqrt{P_{\mathrm{rx}}}, \tag{6.4}$$

where P_{rx} represents the (average) received signal power and $\mathbf{v}_k \sim \mathcal{CN}(\mathbf{0}, \mathbf{I})$ is independent for all k, i.e., Rayleigh fading is assumed for small-scale fading. This power control policy is known as statistics-aware power control, which is easier to be implemented as it only needs to know the distance between the device and the BS. Since host devices may meet local scatterers and obstacles, we assume that $\mathbf{h}_k$ represents a Rayleigh fading channel, which can be expressed as

$$\mathbf{h}_k = \sqrt{\mu_0 d_{ub}^{-\alpha}}\mathbf{v}_k = \sqrt{\frac{\mu_0}{\left\|\mathbf{U}^k - \mathbf{B}\right\|^{\alpha}}}\mathbf{v}_k, \tag{6.5}$$

where d_{ub} is the distance between the host device and the BS, μ_0 is the reference channel gain, and α is the path-loss exponent.

A2) A symmetric signal constellation is used for modulation. As a result, we have $\mathbb{E}[\mathbf{a}_k] = \mathbf{0}$ and $\mathbb{E}[\mathbf{a}_k\mathbf{a}_k^{\mathrm{H}}] = \mathbf{I}$, i.e., the symbol energy is normalized.

Then, the BS can use the correlator to estimate the channel vector, which is shown as

$$\begin{aligned}\tilde{\mathbf{g}}_l &= \mathbf{Y}\mathbf{c}_l = \sum_{k=1}^{K}\mathbf{v}_k\sqrt{P_{\mathrm{rx}}}\delta_{l(k),l} + \mathbf{N}\mathbf{c}_l \\ &= \sum_{k\in\mathcal{K}_l}\mathbf{v}_k\sqrt{P_{\mathrm{rx}}} + \mathbf{N}\mathbf{c}_l,\end{aligned} \tag{6.6}$$

where $\delta_{l,l'}$ is the Kronecker delta (i.e., $\delta_{l,l'} = 1$ if $l = l'$, and 0 otherwise) and $\mathcal{K}_l$ represents the index set for active host devices that choose preamble l. If the kth active host device is the only device that chooses preamble l, i.e., $l(k) = l$ or $\mathcal{K}_l$, thanks to the orthogonality of preambles, it can be shown that

$$\tilde{\mathbf{g}}_l = \mathbf{v}_k\sqrt{P_{\mathrm{rx}}} + \mathbf{n}_l, \tag{6.7}$$

where $\mathbf{n}_l = \mathbf{N}\mathbf{c}_l \sim \mathcal{CN}\left(\mathbf{0}, N_0\mathbf{I}\right)$.

In order to decode the data packet from the kth active host device, conjugate beamforming can be applied and the output of the beamformer can be given by

$$\begin{aligned}\mathbf{x}_l &= \tilde{\mathbf{g}}_l^{\mathrm{H}}\mathbf{Z} \\ &= \tilde{\mathbf{g}}_l^{\mathrm{H}}\mathbf{v}_k\sqrt{P_{\mathrm{rx}}}\mathbf{a}_k + \sum_{k'\neq k}\tilde{\mathbf{g}}_l^{\mathrm{H}}\mathbf{v}_{k'}\sqrt{P_{\mathrm{rx}}}\mathbf{a}_{k'}^{\mathrm{T}} + \tilde{\mathbf{g}}_l^{\mathrm{H}} + \tilde{\mathbf{N}}.\end{aligned} \tag{6.8}$$

Therefore, if $|\tilde{\mathbf{g}}_l^{\mathrm{H}}\mathbf{v}_k|^2$ is sufficiently larger than $|\tilde{\mathbf{g}}_l^{\mathrm{H}}\mathbf{v}_{k'}|^2$, the data packet from the kth active host device can be successfully decoded.

2) **Grant-free RA with parasite devices:** As mentioned earlier, suppose that a host device can serve a number of parasite devices, where parasite devices can be power-limited devices. We assume that when a host device does not have data to send, it can help a parasite device with data to send to the BS.

For convenience, we ignore the index of parasite devices in this part. Then, as illustrated in Fig. 6.2, the received signal at the BS during the preamble transmission phase can be re-written as

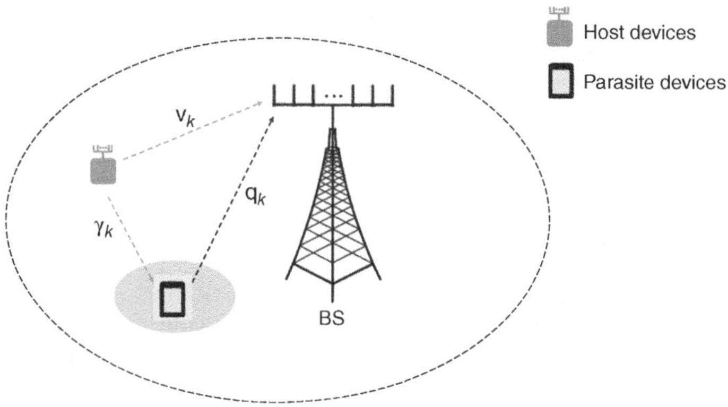

Figure 6.2 Ambient backscatter communications for parasite devices.

$$\mathbf{Y} = \sum_{k=1}^{K} (\mathbf{h}_k + \gamma_k v_k \mathbf{g}_k)\sqrt{P_{\rm k}}\mathbf{c}_{l(k)}^{\rm T} + \mathbf{N}$$
$$= \sum_{k=1}^{K} (\mathbf{v}_k + \gamma_k v_k \mathbf{q}_k)\sqrt{P_{\rm rx}}\mathbf{c}_{l(k)}^{\rm T} + \mathbf{N}. \tag{6.9}$$

Here, $v_k \in \{0, 1\}$ is the activity variable for the kth parasite device, which becomes 1 if the parasite device has data to send, or 0 if it does not, γ_k is the reflection coefficient of the parasite device corresponding to the kth host device, $\mathbf{q}_k$ denotes the small-scale fading channel vector of the parasite device (to the BS) associated with the kth host device, which is referred to as the kth host device-aided parasite device, and $\mathbf{g}_k$ represents the Rayleigh fading channel from the parasite device to the BS, which can be expressed as

$$\mathbf{g}_k = \sqrt{\mu_0 d_{pb}^{-\alpha}}\mathbf{q}_k = \sqrt{\frac{\mu_0}{||\mathbf{P} - \mathbf{B}||^{\alpha}}}\mathbf{q}_k, \tag{6.10}$$

where d_{pb} is the distance between the parasite device and the BS. Note that the BS does not know whether or not the kth host device's signal is used as an ambient signal for its parasite device. Thus, v_k is unknown to the BS.

The channel estimate for the device that uses the lth preamble is given by

$$\tilde{\mathbf{g}}_l = \mathbf{Y}\mathbf{c}_l = \sum_{k \in \mathcal{K}_l} (\mathbf{v}_k + \gamma_k v_k \mathbf{q}_k)\sqrt{P_{\rm rx}} + \mathbf{n}_l$$
$$= (\mathbf{h}_k + \gamma_k v_k \mathbf{g}_k)\sqrt{P_k} + \mathbf{n}_l, \tag{6.11}$$

where the last equality is valid if $\mathcal{K}_l = \{k\}$.

It is assumed that when a host device helps its parasite devices, it transmits a void signal, which is fixed and known to the BS. In particular, let the void signal be an unmodulated signal, which is $\mathbf{1} = [1 \ldots 1]^{\mathrm{T}}$. With parasite devices, the received signal during the data transmission phase is re-written as

$$\mathbf{Z} = \sqrt{P_{\mathrm{rx}}} \sum_{k=1}^{K} \boldsymbol{\Psi}_k + \tilde{\mathbf{N}} \in \mathbb{C}^{M \times D}, \tag{6.12}$$

where

$$\boldsymbol{\Psi}_k = \begin{cases} \mathbf{v}_k \mathbf{1}^{\mathrm{T}} + \gamma_k \mathbf{q}_k \mathbf{b}_k^{\mathrm{T}}, & \text{if } \nu_k = 1, \\ \mathbf{v}_k \mathbf{a}_k^{\mathrm{T}}, & \text{if } \nu_k = 0. \end{cases} \tag{6.13}$$

Here, $\mathbf{b}_k$ represents the data block transmitted from the parasite device corresponding to the kth host device. For backscatter communications, on–off keying (OOK) is used by parasite devices and we have $[\mathbf{b}_k]_d \in \{0, 1\}$ in this case. It is also assumed that $[\mathbf{b}_k]_d$ is independent and

$$\Pr([\mathbf{b}_k]_d = 0) = \Pr([\mathbf{b}_k]_d = 1) = \frac{1}{2}. \tag{6.14}$$

In addition, there are a few remarks as follows:

- A parasite device can send its data to the associated host device so that the host device can forward the data as a relay node to the BS. However, in terms of data privacy, this approach may not be desirable, since the information of the parasite device is known to its host device.
- Furthermore, in backscatter communications, it is not necessary to use a relay protocol between parasite devices and host devices, as the parasite node sends its data directly to the BS.

6.2 Central Limit Theorem-based Signal Detection

In the backscatter communication system, when parasite devices transmit data, the host devices transmit a void signal instead of data. After data transmissions by host devices and parasite devices, the BS needs to perform the following operations [97, 98].

1. **Channel estimation**: At the BS, a bank of correlators with preamble sequences are used to estimate the channel vector associated with each preamble as shown in Eq. (6.11).
2. **Detection of preamble collision**: As for the preamble collision detection, an elementary HOCs-based automatic modulation classification (AMC) approach is introduced in Chapter 4.
3. **Detection of parasite devices**: Since the variables $\{\nu_k\}$ are unknown to the BS, hypothesis testing is to be carried out at the BS with $\mathbf{Z}$ in Eq. (6.12) for the decision of ν_k.

4. **Detection of signals**: If $\nu_k = 0$, a conventional approach can be used to detect the signals from the kth host device. On the other hand, if $\nu_k = 1$, it is necessary to use central limit theorem-based approach to detect signals.

In the following, we show signal detection methods in backscatter systems.

6.2.1 Activity Detection Algorithm

For convenience, we assume that $\mathcal{K}_l = \{k\}$ when detecting the presence of a generalized parasite device's signal in the output of the lth correlator. In addition, let $P = P_{\rm rx}$ and assume that the host device k is the device of interest.

1) **Hypothesis testing:** As indicated in Eq. (6.12), the interference-plus-noise at the mth antenna of the BS during the dth symbol interval becomes

$$w_{k;m,d} = \sqrt{P} \sum_{k' \neq k} I_{k';m,d} + \tilde{n}_{m,d}, \tag{6.15}$$

where

$$I_{k;m,d} = (1 - \nu_k) \upsilon_{k,m} a_{k,d} + \nu_k (\upsilon_{k,m} + \gamma_k q_{k,m} b_{k,d}). \tag{6.16}$$

Let $\sigma^2_{I;m,d}$ and $\sigma^2_{w;m,d}$ denote the variance of $I_{k;m,d}$ and the variance of $w_{k;m,d}$, respectively. Then, we have

$$\sigma^2_{w;m,d} = P(K-1)\sigma^2_{I;m,d} + N_0. \tag{6.17}$$

For tractable analysis, we assume that $w_{k;m,d}$ is a circularly symmetric complex Gaussian (CSCG) random variable.

For convenience, we omit the index k. Then, the hypothesis test can be given by

$$\mathcal{H}_0 : \begin{cases} \tilde{g}_m &= \sqrt{P}\upsilon_m + n_m \\ z_{m,d} &= \sqrt{P}\upsilon_m a_d + w_{m,d}; \end{cases} \tag{6.18a}$$

$$\mathcal{H}_1 : \begin{cases} \tilde{g}_m &= \sqrt{P}(\upsilon_m + \gamma q_m) + n_m \\ z_{m,d} &= \sqrt{P}(\upsilon_m + \gamma q_m b_d) + w_{m,d}. \end{cases} \tag{6.18b}$$

Under $\mathcal{H}_0$, the likelihood function of υ_m is given by

$$f_0(\{\tilde{g}_m, \mathbf{z}_m\} \mid \upsilon_m) = f_0(\tilde{g}_m \mid \upsilon_m) \prod_{d=1}^{D} f_0(z_{m,d} \mid \upsilon_m), \tag{6.19}$$

where $f_0(\tilde{g}_m \mid \upsilon_m) = \mathcal{CN}(\sqrt{P}\upsilon_m, N_0)$ and

$$f_0(z_{m,d} \mid \upsilon_m) = \frac{1}{\pi\sigma_w^2} \prod_{a_d \in \mathcal{A}} e^{-\frac{1}{N_0}|z_{m,d} - \sqrt{P}\upsilon_m a_d|^2}. \tag{6.20}$$

Here, $\mathcal{A}$ denotes the data symbol constellation.

Under $\mathcal{H}_1$, the likelihood function of v_m and γq_m is given by

$$f_1(\{\tilde{g}_m, \mathbf{z}_m\} \mid v_m, \gamma q_m) = f_1(\tilde{g}_m \mid v_m, \gamma q_m) \times \prod_{d=1}^{D} f_1(z_{m,d} \mid v_m, \gamma q_m), \tag{6.21}$$

where $f_1(\tilde{g}_m \mid v_m, \gamma q_m) = \mathcal{CN}(\sqrt{P}(v_m + \gamma q_m), N_0)$ and

$$f_1(z_{m,d} \mid v_m, \gamma q_m) = \frac{1}{\pi\sigma_w^2}\left(e^{-\frac{1}{\sigma_w^2}|z_{m,d}-\sqrt{P}(v_m+\gamma q_m)|^2} + e^{-\frac{1}{\sigma_w^2}|z_{m,d}-\sqrt{P}v_m|^2}\right). \tag{6.22}$$

Then, the generalized likelihood ratio test (GLRT) can be carried out using the test statistic as

$$\Lambda = \frac{\prod_m \max_{v_m} f_0(\{\tilde{g}_m, \mathbf{z}_m\} \mid v_m)}{\prod_m \max_{v_m, \gamma q_m} f_1(\{\tilde{g}_m, \mathbf{z}_m\} \mid v_m, \gamma q_m)}. \tag{6.23}$$

Unfortunately, although it is possible to form the GLRT as in Eq. (6.23), it is difficult to carry out the optimization with $f_1(\{\tilde{g}_m, \mathbf{z}_m\} \mid v_m, \gamma q_m)$. Thus, we need to resort to a suboptimal approach.

2) **A suboptimal approach based on the Central Limit Theorem (CLT):** To efficiently detect the presence of parasite devices, the detection statistic $\alpha_{m,d}$ is proposed by considering the differences of channel information and data together with the correlation between host devices and parasite devices, where $\tilde{g}_m$ and $z_{m,d}$ are used. Thus, let

$$\alpha_{m,d} = \tilde{g}_m^* z_{m,d}. \tag{6.24}$$

Moreover, the decision statistic is considered, which is given as

$$B = \frac{1}{MD}\sum_{d=1}^{D}\sum_{m=1}^{M} \tilde{g}_m^* z_{m,d} = \frac{1}{MD}\sum_{d=1}^{D}\sum_{m=1}^{M} \alpha_{m,d}. \tag{6.25}$$

In order to characterize $\alpha_{m,d}$, we need to consider the relationship between v_m and q_m. As q_m is also a small-scale fading channel coefficient, according to the assumption of **A1**, it can be assumed that $\mathbf{q} \sim \mathcal{CN}(\mathbf{0}, \mathbf{I})$, while q_m can be correlated with v_m due to a short distance between the host device and its parasite devices. Thus, we consider the following assumption.

A3) The pair of (v_m, q_m) is independent of other pairs, $(v_{m'}, q_{m'})$, where $m' \neq m$, and it is given by

$$\begin{bmatrix} v_m \\ q_m \end{bmatrix} \sim \mathcal{CN}\left(\mathbf{0}, \begin{bmatrix} 1 & \rho \\ \rho^* & 1 \end{bmatrix}\right), \tag{6.26}$$

where $\rho = \mathbb{E}[v_m q_m^*]$.

Under $\mathcal{H}_0$, the statistical expectation and variance of $\alpha_{m,d}$ can be given by

$$\mathbb{E}[\alpha_{m,d} \mid \mathcal{H}_0] = 0, \tag{6.27}$$

and

$$\begin{aligned}\mathrm{Var}(\alpha_{m,d} \mid \mathcal{H}_0) &= P^2 \left(\mathbb{E}[|v_m|^4] + \mathbb{E}[|v_m|^2]^2 \,\mathrm{Var}(a_d)\right) \\ &\quad + P(N_0 + \sigma^2_{w;m,d})\mathbb{E}[|v_m|^2] + N_0\sigma^2_{w;m,d},\end{aligned} \tag{6.28}$$

respectively, under the assumption of **A2**. According to the assumption of **A1**, we have $\mathbb{E}[|v_m|^2] = 1$. In addition, since $|v_m|^2$ has an exponential distribution, it can be shown that $\mathbb{E}[|v_m|^4] = 2$. Thus, we have

$$\mathrm{Var}(\alpha_{m,d} \mid \mathcal{H}_0) = 3P^2 + P\left(\sigma^2_{w;m,d} + N_0\right) + N_0\sigma^2_{w;m,d}, \tag{6.29}$$

where

$$\sigma^2_{w;m,d} = P(K-1)\left(1 + \frac{\gamma^2}{4} + \frac{\gamma\rho}{2}\right) + N_0. \tag{6.30}$$

Based on the CLT for a large M and D, the decision statistic B_0 can be approximated using a Gaussian random variable, i.e.,

$$B_0 \sim \mathcal{CN}(0, \sigma_0^2). \tag{6.31}$$

Let $\sigma^2_{w;m}$ denote the variance of $\sum_{d=1}^{D} w_{k;m,d}$. σ_0^2 can be calculated by

$$\sigma_0^2 = \frac{1}{M}\left(\frac{1}{D}2P^2 + P\sigma^2_{w;m} + \frac{1}{D}PN_0 + N_0\sigma^2_{w;m}\right) + \frac{P^2}{D}. \tag{6.32}$$

Since not all the variables are independent and identically distributed to d, we consider the distribution of each variable. Then, $\sigma^2_{w;m}$ can be written as

$$\sigma^2_{w;m} = P(K-1)\left[\frac{1}{2} + \frac{5}{2D} + \frac{\gamma^2}{8D} + \frac{\gamma^2}{8} + \frac{\gamma\rho}{4}\right] + \frac{N_0}{D}. \tag{6.33}$$

Under $\mathcal{H}_1$, it can be shown that

$$\mathbb{E}[\alpha_{m,d} \mid \mathcal{H}_1] = P\left(1 + \gamma^*\rho + \frac{\gamma\rho^*}{2} + \frac{|\gamma|^2}{2}\right), \tag{6.34}$$

and

$$\begin{aligned}\mathrm{Var}(\alpha_{m,d} \mid \mathcal{H}_1) &= P^2\left[2 + \gamma^2\left(\frac{3}{2} + \frac{7\rho^2}{4} + \frac{5\gamma^2}{4}\right)\right] \\ &\quad + P\left[\left(1 + \frac{\gamma^2}{2}\right)N_0 + \left(1 + 2\gamma\rho + \gamma^2\right)\sigma^2_{w;m,d}\right] \\ &\quad + P^2\left[4\gamma\rho + \gamma^2(1 + 4\rho + 3\rho^2) + \gamma^3\rho\right] \\ &\quad + N_0\sigma^2_{w;m,d}.\end{aligned} \tag{6.35}$$

Therefore, the decision statistic under $\mathcal{H}_1$ can be shown as

$$B_1 \sim \mathcal{CN}(\mu_1, \sigma_1^2), \tag{6.36}$$

where

$$\mu_1 = \mathbb{E}[\alpha_{m,d} \mid \mathcal{H}_1], \tag{6.37}$$

and

$$\begin{aligned}\sigma_1^2 = &\frac{P^2}{M}\left\{2 + \frac{\gamma^2}{4D}\left[(5D+1)(1+\rho^2) + \rho^2 M\right] + \frac{\gamma^4}{4}\left(1 + \frac{2+M}{D}\right)\right\}\\ &+ \frac{P}{M}\left[(1+2\gamma\rho+\gamma^2)\sigma_{w;m}^2 + N_0 + \gamma^2\left(\frac{N_0}{4D} + \frac{N_0}{4}\right)\right]\\ &+ \frac{1}{M}\left\{N_0\sigma_{w;m}^2 + P^2\left[6\gamma\rho + \gamma^2(1+2\rho+3\rho^2) + \gamma^3\rho\right]\right\}.\end{aligned} \tag{6.38}$$

Finally, the maximum likelihood (ML) detection can be used to detect the presence of parasite devices, which is

$$\ln \frac{f(B \mid \mathcal{H}_0)}{f(B \mid \mathcal{H}_1)} \underset{\mathcal{H}_0}{\overset{\mathcal{H}_1}{\lessgtr}} 0, \tag{6.39}$$

where

$$f(B \mid \mathcal{H}_0) = \frac{1}{\sqrt{2\pi\sigma_0^2}} \exp\left(-\frac{1}{\sigma_0^2}B^2\right), \tag{6.40}$$

$$f(B \mid \mathcal{H}_1) = \frac{1}{\sqrt{2\pi\sigma_1^2}} \exp\left[-\frac{1}{\sigma_1^2}(B-\mu_1)^2\right]. \tag{6.41}$$

6.2.2 Signal Detection Algorithm

In this section, we discuss the signal detection of parasite devices when $\nu_k = 1$. Let

$$G_p = \frac{1}{M}\sum_m |\tilde{g}_m|^2, \tag{6.41}$$

and

$$G_d = \frac{1}{M}\sum_m \alpha_{m,d}. \tag{6.42}$$

Then, the test statistic for the detection of parasite devices' signal is

$$G = G_p - G_d, \tag{6.43}$$

which can be approximated by a Gaussian random variable.

If $b_d = 0$, we have

$$\mathbb{E}[G \mid b_d = 0] = P(\gamma^2 + \gamma\rho^*), \tag{6.44}$$

and

$$\begin{aligned}\mathrm{Var}(G \mid b_d = 0) = &\frac{1}{M}\left[P(N_0 + \sigma^2_{w;m,d})(1 + \gamma^2 + 2\gamma^*\rho) + N_0(2N_0 + \sigma^2_{w;m,d})\right] \\ &+ \frac{1}{M}\left[P\gamma^2(1 + 2\gamma^2 + \rho^2) + \gamma^2 N_0\right] + \frac{1}{M}2\sqrt{P}(\gamma^2 + \gamma\rho^*)N_0.\end{aligned} \tag{6.45}$$

If $b_d = 1$, it can be shown that

$$\mathbb{E}[G \mid b_d = 1] = N_0, \tag{6.46}$$

and

$$\mathrm{Var}(G \mid b_d = 1) = \frac{1}{M}\left\{P(N_0 + \sigma^2_{w;m,d})(1 + \gamma^2 + 2\gamma^*\rho) + N_0(2N_0 + \sigma^2_{w;m,d})\right\}. \tag{6.47}$$

Therefore, the ML detection can be shown as

$$\ln \frac{f(G \mid b_d = 0)}{f(G \mid b_d = 1)} \underset{b_d=0}{\overset{b_d=1}{\lessgtr}} 0. \tag{6.48}$$

Note that the channel estimation error affects the variance of decision statistic as less channel estimation error results in a smaller variance, which is easier to differentiate host signals and parasite signals.

6.2.3 Performance Analysis

In the following, we analyze the detection performance of the presence of parasite devices and the bit error rate (BER) of the signal detection.

Firstly, we define the probability of error detection, denoted by P_e, in the presence detection as total probability of an incorrect decision, which consists of the probability that host device's data signal is misjudged to the combination of parasite device's data signal and host device's void signal, and vice versa. Therefore, the probability of error detection of the presence detection in Eq. (6.39) is given by

$$\begin{aligned}P_e &= \Pr(v_k = 0)P_{e\mid 0} + \Pr(v_k = 1)P_{e\mid 1} \\ &= \frac{1}{2\sqrt{2\pi\sigma_0^2}}\int_{\delta_e}^{\infty} \exp\left(-\frac{(x - \mu_0)^2}{2\sigma_0^2}\right) dx\end{aligned}$$

$$+\frac{1}{2\sqrt{2\pi\sigma_1^2}}\int_{-\infty}^{-\delta_e}\exp\left(-\frac{(x-\mu_1)^2}{2\sigma_1^2}\right)dx$$
$$=\frac{1}{2}Q\left(\frac{\mu_1-\delta_e}{\sqrt{\sigma_1^2}}\right)+\frac{1}{2}Q\left(\frac{\delta_e-\mu_0}{\sqrt{\sigma_0^2}}\right), \tag{6.49}$$

where $\Pr(v_k=0)$ and $\Pr(v_k=1)$ denote the probability that the kth parasite device is inactive and the kth parasite device is active, respectively, $P_{e\,|\,0}$ and $P_{e\,|\,1}$ denote the probability of the condition that the host device is misjudged to the parasite device and vice versa, respectively, and δ_e can be written as:

$$\delta_e=\frac{\mu_1\sigma_0^2-\sqrt{\mu_1^2\sigma_0^4-(\sigma_0^2-\sigma_1^2)\left(\mu_1^2\sigma_0^2-2\sigma_0^2\sigma_1^2\ln\left(\frac{\sigma_0}{\sigma_1}\right)\right)}}{\sigma_0^2-\sigma_1^2}. \tag{6.50}$$

Furthermore, for convenience, let $\mu_{b0}=\mathbb{E}[G\,|\,b_d=0]$, $\sigma_{b0}^2=\mathrm{Var}(G\,|\,b_d=0)$, $\mu_{b1}=\mathbb{E}[G\,|\,b_d=1]$, and $\sigma_{b1}^2=\mathrm{Var}(G\,|\,b_d=1)$. The expression of BER of the signal detection is shown as follows:

$$P=\Pr([\mathbf{b}_k]_d=0)P_{e\,|\,0}+\Pr([\mathbf{b}_k]_d=1)P_{e\,|\,1}$$
$$=\frac{1}{2}Q\left(\frac{\mu_{b0}-\delta_d}{\sqrt{\sigma_{b0}^2}}\right)+\frac{1}{2}Q\left(\frac{\delta_d-\mu_{b1}}{\sqrt{\sigma_{b1}^2}}\right), \tag{6.51}$$

where $\Pr([\mathbf{b}_k]_d=0)$ and $\Pr([\mathbf{b}_k]_d=1)$ denote the probability that the dth symbol of parasite signal is 0 and the dth symbol of parasite signal is 1, respectively, and δ_d can be written as

$$\delta_d=\frac{\mu_{b1}\sigma_{b0}^2-\mu_{b0}\sigma_{b1}^2}{\sigma_{b0}^2-\sigma_{b1}^2}$$
$$-\frac{\sqrt{\left(\mu_{b1}\sigma_{b0}^2-\mu_{b0}\sigma_{b1}^2\right)^2-\left(\sigma_{b0}^2-\sigma_{b1}^2\right)\left(\mu_{b1}^2\sigma_{b0}^2-\mu_{b0}^2\sigma_{b1}^2-2\sigma_{b0}^2\sigma_{b1}^2\ln\left(\frac{\sigma_{b0}}{\sigma_{b1}}\right)\right)}}{\sigma_{b0}^2-\sigma_{b1}^2}. \tag{6.52}$$

6.3 Simulation Results

We assume that there are K host devices in the backscatter communication system and each host device is to help I parasite devices. The background noise is additive white Gaussian noise (AWGN) and the signal-to-noise ratio (SNR) is defined as

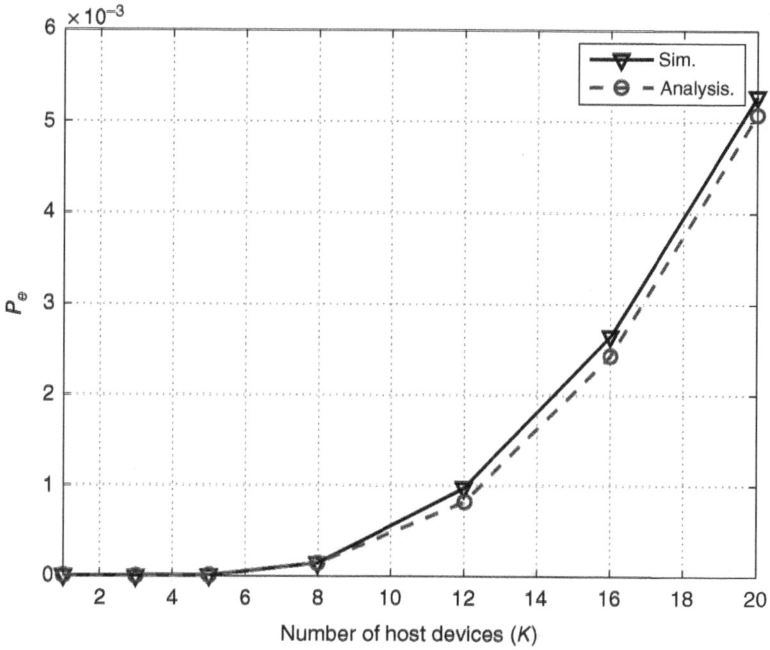

Figure 6.3 P_e versus number of host devices, K, with $M = 200$, SNR = 15 dB, $D = 100$, $L = 64$, $\rho = 0.4$, and $\gamma = 1$.

the ratio of the received signal at the BS to the background noise. Note that the multiple-preamble scheme can be used to ensure partial orthogonality between preambles and greatly reduces the probability of collision [99]. Thus, we assume that no collisions happen in our system.

Firstly, we show the probability of error detection in terms of the number of host devices, K, the number of antennas, M, SNR, and the channel correlation coefficient, ρ, to verify the effectiveness of detection method. In Fig. 6.3, the probability of error detection of the CLT-based approach for different numbers of hose devices, K, is shown when $M = 200$, SNR = 15 dB, $D = 100$, $L = 64$, $\rho = 0.4$, and $\gamma = 1$ [100]. It is shown that the probability of error detection based on CLT increases as the number of host devices increases because the interference between signals of different devices increases when there are more host devices. Thus, it becomes more difficult to differentiate between host devices and parasite devices.

Figure 6.4 shows the probability of error detection of the CLT-based approach with SNR when $M = 200$, $D = 100$, $L = 64$, $\rho = 0.4$, $\gamma = 1$, and $K = 5, 10, 15$. We notice that the detection performance improves as SNR increases, which indicates that at a sufficiently high SNR, the BS can carry out successful detection.

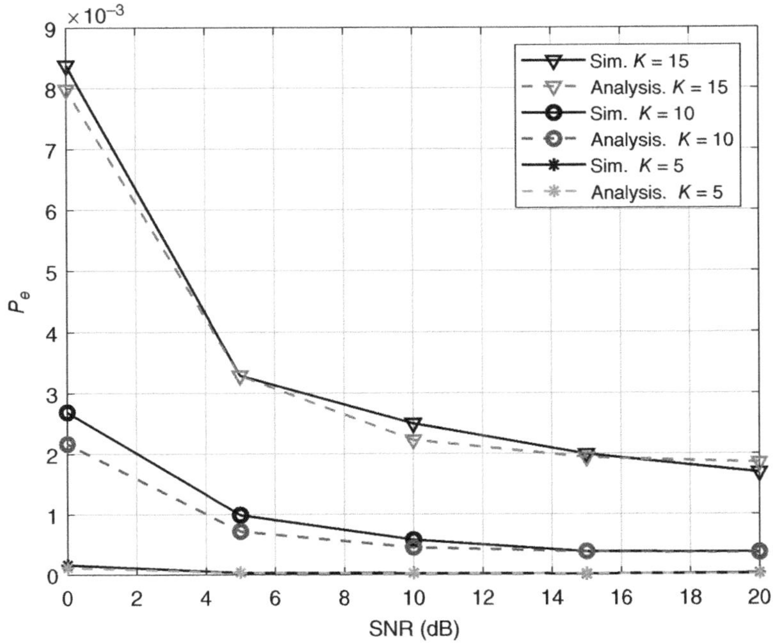

Figure 6.4 P_e versus SNR, with $M = 200$, $D = 100$, $L = 64$, $\rho = 0.4$, and $\gamma = 1$.

It also indicates that the CLT-based approach provides a good presence detection performance when the number of host devices is up to 15.

Figure 6.5 shows the probability of error detection of the CLT-based approach with different numbers of antennas, M, when SNR $= 15$ dB, $D = 100$, $L = 64$, $\rho = 0.4$, $\gamma = 1$, and $K = 5, 10, 15$. From Fig. 6.5, we can see that the probability of error detection decreases as the number of antennas M increases and the CLT-based approach can provide a good presence detection performance. Since the advantage of a large number of antennas can be taken to eliminate the effects of small-scale fading, the variance decreases as the number of antennas increases.

Figure 6.6 shows the probability of error detection of the CLT-based approach with the channel correlation coefficient, ρ, when $M = 200$, SNR $= 15$ dB, $D = 100$, $L = 64$, $\gamma = 1$, and $K = 10, 15$. It can be seen that the probability of error detection decreases rapidly, and then tends to be steady as the channel correlation coefficient ρ increases. Since host devices transmit a void signal instead of data when parasite devices transmit data, it is easier to detect the presence of parasite signals from received signals through the energy of detection statistics when the channel state information (CSI) of parasite and host devices is more relevant.

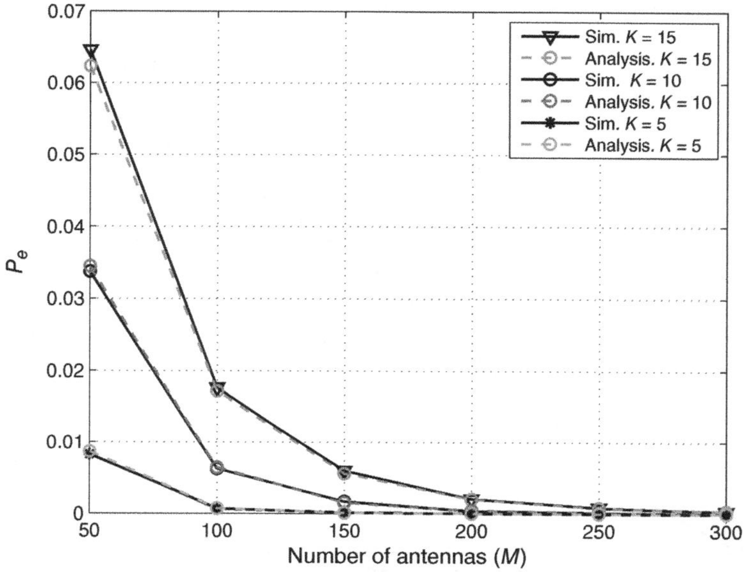

Figure 6.5 P_e versus number of antennas, M, with SNR = 15 dB, $D = 100$, $L = 64$, $\rho = 0.4$, and $\gamma = 1$.

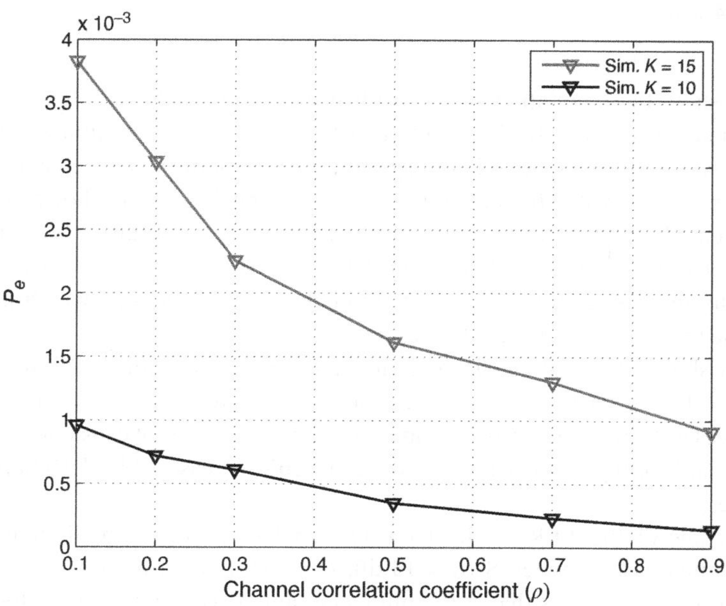

Figure 6.6 P_e versus channel correlation coefficient, ρ, with $M = 200$, SNR = 15 dB, $D = 100$, $L = 64$, and $\gamma = 1$.

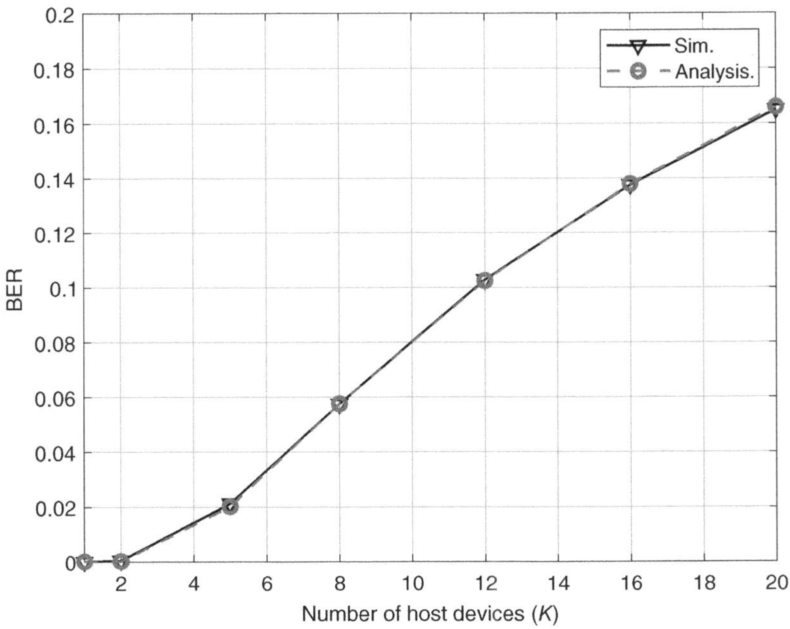

Figure 6.7 BER versus number of host devices, K, with $M = 200$, SNR $= 15$ dB, $D = 100$, $L = 64$, $\rho = 0.4$, and $\gamma = 1$.

Then, we show the simulation results with respect to the BER to illustrate the performance of signal detection from parasite devices. Since the number of host devices is an important factor in our system, the BER of the signal detection from parasite devices with different numbers of host devices, K, is shown in Fig. 6.7, when $M = 200$, SNR $= 15$ dB, $D = 100$, $L = 64$, $\rho = 0.4$, and $\gamma = 1$. It is shown that the numerical and theoretical BER of CLT-based approach are monotonically increasing functions of the number of host devices K, which is caused by the interference between devices.

Figure 6.8 shows the BER of the parasite signal detection with SNR when $M = 200$, $D = 100$, $L = 64$, $\rho = 0.4$, $\gamma = 1$, and $K = 2, 5, 10, 15$. It is observed that the curves of BER decline rapidly and then tend to be stable. We also notice that with a sufficiently high SNR, the BS can achieve reasonable detection performance.

Figure 6.9 shows the BER of the parasite signal detection with different numbers of antennas, M, when SNR $= 15$ dB, $D = 100$, $L = 64$, $\rho = 0.4$, $\gamma = 1$, and $K = 2, 5, 10, 15$. It shows that as the number of antennas increases, the BER curves decline in a similar trend, where the curve with a smaller K has lower BER in general. It is noteworthy that more antennas can reduce the error of channel

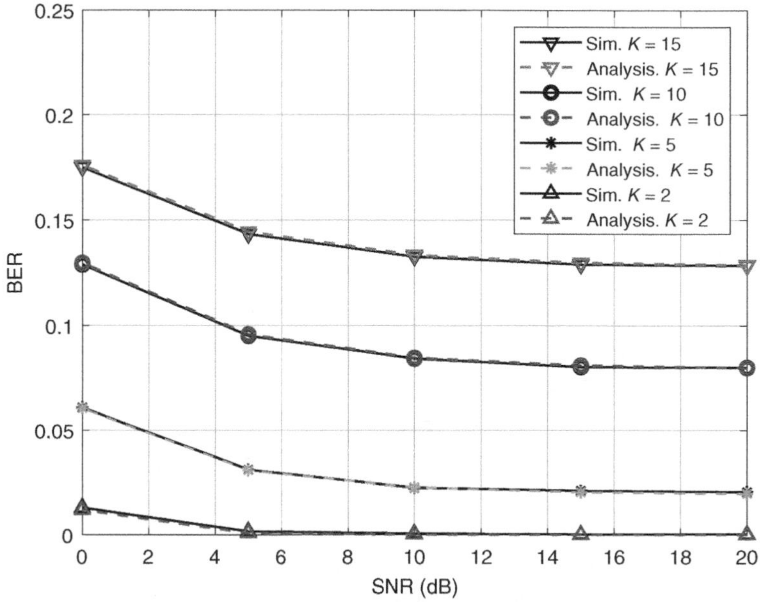

Figure 6.8 BER versus SNR, with $M = 200$, $D = 100$, $L = 64$, $\rho = 0.4$, and $\gamma = 1$.

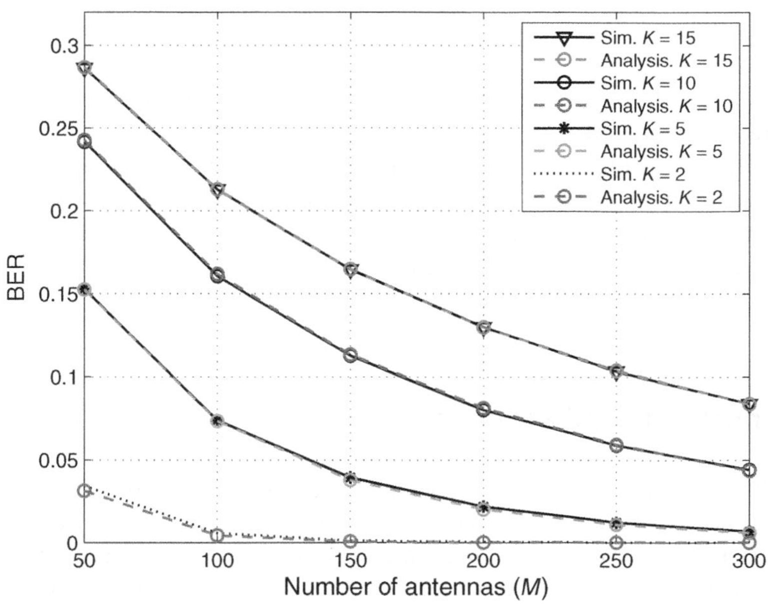

Figure 6.9 BER versus number of antennas, M, with SNR = 15 dB, $D = 100$, $L = 64$, $\rho = 0.4$, and $\gamma = 1$.

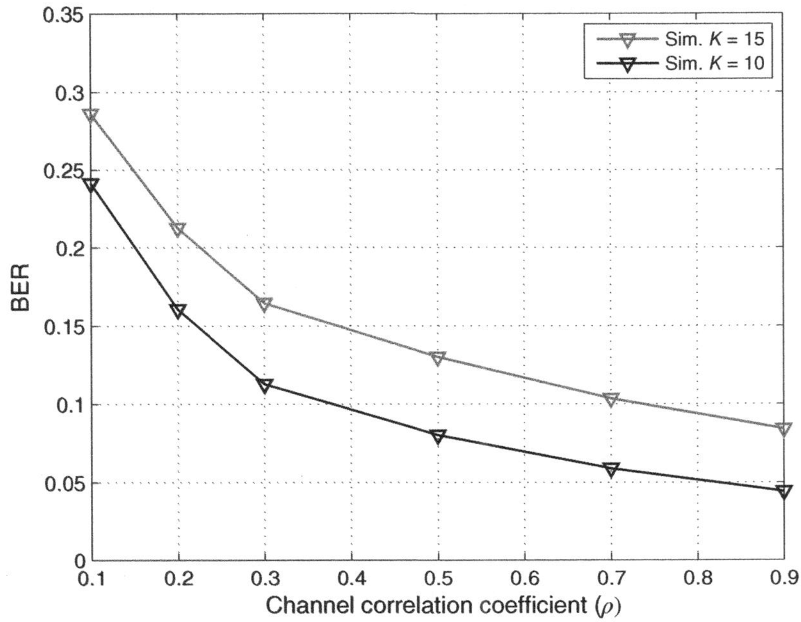

Figure 6.10 BER versus channel correlation coefficient, ρ, with SNR $= 15\,dB$, $M = 200$, $D = 100$, $L = 64$, $K = 10$, and $\gamma = 1$.

estimation and make the CLT-based approach better fit the Gaussian distribution, thereby reducing the BER.

Figure 6.10 shows the BER of the parasite signal detection with channel correlation coefficient, ρ, when $M = 200$, SNR $= 15$ dB, $D = 100$, $L = 64$, $\gamma = 1$, and $K = 10$, 15. It is indicted that the BER decreases as the channel correlation coefficient ρ increases and the P_e in the presence detection affects the performance in the signal detection.

6.4 Conclusion and Remarks

In this chapter, a backscatter communication system has been proposed, where parasite devices were able to use the transmitted signals from host devices as ambient signals for backscatter communications and send data packets to a BS. Since massive MIMO has been used at the BS, a CLT-based approach was introduced to detect the presence of parasite devices and separate parasite signals from host signals.

Backscatter communications allow parasite devices use ambient signals of host devices to send their data packets, while the technique of non-orthogonal multiple access (NOMA) allows multiple devices to occupy the same frequency, time, or code resource. In Chapter 7, we will introduce the NOMA signal in IoT and investigate the optimization problem of NOMA systems.

7

Analysis and Optimization for NOMA Signals

- The key concept of non-orthogonal multiple access (NOMA) is that of allowing multiple Internet of Things (IoT) devices to occupy the same frequency-, time-, or code-resource [101, 102]. The device having better channel conditions performs successive interference cancellation (SIC) [103]. As a remedy, full-duplex (FD) can be invoked by cooperative NOMA to avoid halving the rate by conventional half-duplex (HD), where the strong user (SU) with good channel conditions can act as an FD relay node for the weak user (WU) with poor channel conditions [104–106]. However, the energy efficiency (EE) of cooperative NOMA may be degraded due to additional power consumption incurred at the SUs [107–109]. In this chapter, we investigate the EE maximization problem of an FD cooperative NOMA system. More importantly, we introduce the signal-to-interference-plus-noise ratio (SINR) gap reversal problem of cooperative NOMA systems, which imposes SIC performance degradation at the SUs.

7.1 System Model

We consider a NOMA system supporting two IoT devices by a base stations (BS) as shown in Fig. 7.1. The IoT users have different channel conditions on the same frequency, which is a common scenario in NOMA systems [103]. The BS, SU, and WU are equipped with a single antenna for the sake of having a low hardware complexity. In particular, the SU acts as a decode-and-forward relay, which cooperates with the WU. FD operation is enabled at the SU, which can transmit and receive signals at the same time [110]. The channel state information (CSI) is obtained by channel estimation in the training phase based on the channel's reciprocity [108]. Explicitly, during the training phase, the BS estimates the downlink CSI by inferring it from the channel quality indicator feedback received from the SU and the

IoT Signal Detection, First Edition. Rui Han, Jingjing Wang, Lin Bai, and Jianwei Liu.

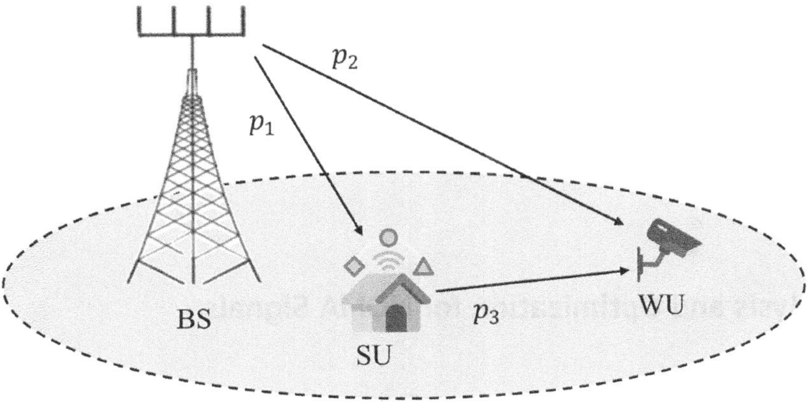

Figure 7.1 NOMA system with two IoT devices.

WU. Hence, the BS can regularly update the CSI of the BS-SU link and the BS-WU link. Furthermore, since the SU is aware of its own channel quality indicator and of the WU, the CSI of the self-interference (SI) leakage channel and the SU-WU link can also be reported to the BS.

Since a three-node single-antenna scenario is assumed, based on the state-of-the-art in overhead research, the signaling overhead (power, bandwidth, and throughput) maintains low [111, 112]. Hence, we only focus our attention on the subsequent transmission stage. Let us define the transmission power allocated to the SU and to the WU at the BS by p_1 and p_2, respectively, where the maximum transmission power available at the BS is upper bounded by $p_{\text{BS,max}}$. In addition, we define the transmission power consumed at the SU by p_3, which is upper bounded by the maximum available transmission power constraint $p_{3,\max}$. Furthermore, to guarantee that the SU can successfully decode the WU's signal and subtract it by SIC, we introduce the constraint of $\Gamma_{12} \geq \Gamma_2$, where Γ_{12} and Γ_2 represent the WU's SINR observed at the SU and the WU's signal at its own end, respectively. In addition, to guarantee the WU's quality-of-service (QoS), we have $\Gamma_2 \geq \Gamma_{2,\text{req}}$, where $\Gamma_{2,\text{req}}$ is the minimum SINR requirement of the WU.

Let us define the EE (in bits/Joule/Hz) as the ratio of the system's throughput T_{total} to the total power consumption P_{total}. To maximize the system's EE, we jointly optimize the transmission power p_1, p_2, and p_3. Accordingly, the EE maximization problem is formulated as

$$\begin{aligned} \text{P1} &: \underset{p_1, p_2, p_3}{\operatorname{argmax}} \frac{T_{\text{total}}}{P_{\text{total}}}, \\ \text{s.t.}\,(\text{C1}) &: p_1 + p_2 \leq p_{\text{BS,max}}, \\ (\text{C2}) &: p_3 \leq p_{3,\max}, \end{aligned}$$

$$(\text{C3}) : 0 \le p_1, 0 \le p_2, 0 \le p_3,$$
$$(\text{C4}) : \Gamma_{12} \ge \Gamma_2,$$
$$(\text{C5}) : \Gamma_2 \ge \Gamma_{2,\text{req}}. \quad (7.1)$$

Constraint (C1) indicates that the transmission power allocated at the BS should be lower than the available transmission power $p_{\text{BS,max}}$. Constraint (C2) means that the transmission power of the SU should be lower than its total available transmission power $p_{3,\text{max}}$. Constraint (C3) implies that all the transmission power should be non-negative. Constraint (C4) guarantees that the SU can decode the WU's signal, while Constraint (C5) suggests that the WU's QoS should be satisfied.

7.2 Throughput and Power Consumption Analysis

The optimization problem involves both throughput and power consumption, which will be introduced in the following, respectively.

7.2.1 Throughput Analysis

At the WU, the received signal arrives both from the BS and the SU, which is given by

$$r_2[i] = (\sqrt{p_1}x_1[i] + \sqrt{p_2}x_2[i])h_2 + \sqrt{p_3}x_2[i-\tau]h_{12} + n[i], \quad (7.2)$$

where $x_1[i]$ and $x_2[i]$ are the ith data symbols intended for the SU and the WU, respectively, $n[i]$ is the additive white Gaussian noise (AWGN), the symbol delay τ is caused by the processing delay at the SU, h_2 is the channel response from the BS to the WU, and h_{12} is the channel response from the SU to the WU. All propagation channels capture the effects of large-scale and small-scale fading. Since the phase of h_2 and h_{12} is different, the desired signal replicas arriving from the BS and the SU are mis-aligned in phase. Hence, the phase of the transmitted signal at the SU has to be shifted to align the pair of received signals. The channel spanning from the BS to the WU can be represented as $h_2 = |h_2|e^{j\theta_2}$, where $|h_2|$ and $e^{j\theta_2}$ are the magnitude and phase of the channel h_2, respectively. Similarly, the channel spanning from the SU to the WU is represented as $h_{12} = |h_{12}|e^{j\theta_{12}}$, where $|h_{12}|$ and $e^{j\theta_{12}}$ are the magnitude and phase of the channel h_{12}, respectively. Therefore, the phase difference between the two channels is calculated as $\phi = \theta_2 - \theta_{12}$. To align the two signals, the signal transmitted from the SU is pre-processed as $(\sqrt{p_3}e^{j\phi})x_2$. As a result, the signal received at the WU becomes

$$r_2[i] = (\sqrt{p_3}|h_{12}|e^{j(\theta_{12}+\phi)}x_2[i-\tau] + \sqrt{p_2}|h_2|e^{j\theta_2}x_2[i]) + \sqrt{p_1}|h_2|e^{j\theta_2}x_1[i] + n[i]. \quad (7.3)$$

As we can see, the pair of desired signals arriving from the BS and the SU are now well aligned at the WU,[1] and the SINR of the WU at its own end is given by

$$\Gamma_2 = \frac{p_2|h_2|^2 + p_3|h_{12}|^2}{p_1|h_2|^2 + \sigma^2}. \tag{7.4}$$

The signal received at the SU is given by

$$r_1[i] = (\sqrt{p_1}x_1[i] + \sqrt{p_2}x_2[i])h_1 + (\sqrt{p_3}e^{j(\phi+\theta_r)}x_2[i-\tau])\frac{h_r}{\sqrt{\alpha}} + n[i], \tag{7.5}$$

where h_1 and h_r represent the channel spanning from the BS to the SU and the self-interference (SI) leakage channel from the SU's transmitter to its receiver, respectively. Still referring to Eq. (7.5), α is the SI reduction factor defined as the ratio of the SI powers before and after SI suppression. Therefore, the WU's SINR observed at the SU is given by

$$\Gamma_{12} = \frac{p_2|h_1|^2}{p_3|\tilde{h}_r|^2 + p_1|h_1|^2 + \sigma^2}, \tag{7.6}$$

where $\tilde{h}_r$ is the residual SI channel. To guarantee successful SIC at the SU, we have to ensure that $\Gamma_{12} \geq \Gamma_2$. Based on Eqs. (7.4) and (7.6), the inequality $\Gamma_{12} \geq \Gamma_2$ is further derived into $\frac{p_2|h_1|^2}{p_3|\tilde{h}_r|^2+p_1|h_1|^2+\sigma^2} \geq \frac{p_2|h_2|^2+p_3|h_{12}|^2}{p_1|h_2|^2+\sigma^2}$. Since $|h_2|^2 \leq |h_1|^2$, we have $\Delta = (p_1|h_1|^2|h_{12}|^2 + |h_{12}|^2\sigma + p_2|h_2|^4)^2 - 4|h_{12}|^2|h_2|^2p_2(|h_2|^2 - |h_1|^2)\sigma^2 \geq 0$, As a result, the feasible domain of p_3 can be represented in the closed form

$$p_3 \in \left[0, \frac{-B+\sqrt{\Delta}}{2A}\right], \tag{7.7}$$

where we have $B = p_1|h_1|^2|h_{12}|^2 + |h_{12}|^2\sigma + p_2|h_2|^4$ and $A = |h_{12}|^2|h_2|^2$. Finally, the SU's SINR after SIC is calculated as

$$\Gamma_1 = \frac{p_1|h_1|^2}{p_3|\tilde{h}_r|^2 + \sigma^2}. \tag{7.8}$$

7.2.2 Power Consumption Analysis

The total power consumption mainly consists of the power amplifier (PA) power, circuit power of the transmit/receive chains, and the power consumed by SI cancellation [113].

a) The PA power is closely related to the radiated transmit power and the drain efficiency of the PA, which is given by $\frac{1}{\eta}(p_1 + p_2 + p_3)$ [113], where η is the drain

1 The signal received at the WU has a low time delay, which can be mitigated by an equalizer or sequence detector. Hereby, we assume that the signal from the BS and the SU can be readily combined by the WU [106].

efficiency of the PA. Without loss of generality, we assume that all PAs of the BS and of the users have the same drain efficiency performance.

b) The circuit power consumption of the transmit/receive chains is proportional to the number of active transmit and receive chains, including the power consumed by the digital-to-analog converter, by the filter and the synthesizer, etc. The dynamic power consumption can be calculated as $2(p_{c,r} + p_{c,t})$, where $p_{c,r}$ and $p_{c,t}$ denote the circuit power consumed by the receive and transmit chains of the BS, of the SU, and of the WU. The multiplier 2 indicates that we have two transmit chains and two receive chains in the system.
c) For the SU acting as an FD relay, additional power p_{SI} is required for SI cancellation, which can be modeled by a constant. Generally speaking, complex SI cancellation schemes dissipate higher power by their digital-to-analog converter, transmit radio unit, and adders [113].
d) The fixed power consumption p_{fix} represents power consumed by the power supply, by the active cooling system, etc. This part is independent of the state of the transmit/receive chains [113]. Finally, the total power consumption is given by

$$P_{total} = \frac{p_1 + p_2 + p_3}{\eta} + 2(p_{c,r} + p_{c,t}) + p_{SI} + p_{fix}. \tag{7.9}$$

7.3 Energy Efficiency Performance Optimization

For simplicity, we collect the transmission power into a vector $\mathbf{p} = [p_1, p_2, p_3]^T$ and define $\{\boldsymbol{\Theta}\}$ as the feasible domain confined by the constraints. Therefore, the EE maximization problem is re-formulated as:

$$\begin{aligned}
\text{P2}: \underset{\mathbf{p}\in\{\boldsymbol{\Theta}\}}{\text{argmax}} & \frac{\log_2(1 + \frac{\mathbf{a}_1\mathbf{p}|h_1|^2}{\mathbf{a}_3\mathbf{p}|\tilde{h}_r|^2+\sigma^2}) + \log_2(1 + \frac{\mathbf{a}_2\mathbf{p}|h_2|^2+\mathbf{a}_3\mathbf{p}|h_{12}|^2}{\mathbf{a}_1\mathbf{p}|h_2|^2+\sigma^2})}{\frac{1}{\eta}(\|\mathbf{p}\|_1) + 2(p_{c,r} + p_{c,t}) + p_{fix} + p_{SI}}, \\
\text{s.t.}\ & (\hat{C}_1): \mathbf{a}_1\mathbf{p} + \mathbf{a}_2\mathbf{p} \leq p_{BS,max}, \\
& (\hat{C}_2): \mathbf{a}_3\mathbf{p} \leq p_{3,max}, \\
& (\hat{C}_3): \mathbf{0} \preceq \mathbf{p}, \\
& (C4): \Gamma_{12} \geq \Gamma_2, \\
& (C5): \Gamma_2 \geq \Gamma_{2,req},
\end{aligned} \tag{7.10}$$

where $\mathbf{a}_1 = [1, 0, 0]$, $\mathbf{a}_2 = [0, 1, 0]$, and $\mathbf{a}_3 = [0, 0, 1]$. The operator $\|\cdot\|_1$ represents the 1-norm of vector. It can been seen that the objective function is non-convex and (C4) (C5) represent quadratic constraints, which impose a challenge in terms of obtaining a globally optimal result within polynomial time. To strike an attractive balance between the performance and complexity, a low-complexity near-optimal solution is desirable. We observe that the numerator of the objective function is the difference between two convex functions. To handle this challenge, we first reformulate the objective function of Eq. (7.10) as $\underset{\mathbf{p}\in\{\mathbf{\Theta}\}}{\operatorname{argmin}} \frac{f(\mathbf{p})-g(\mathbf{p})}{\frac{1}{\eta}\|\mathbf{p}\|_1+2(p_{c,r}+p_{c,t})+p_{fix}+p_{SI}}$, where $f(\mathbf{p}) = -\log_2\left(\mathbf{a}_3\mathbf{p}|\tilde{h}_r|^2 + \mathbf{a}_1\mathbf{p}|h_1|^2 + \sigma^2\right) - \log_2\left(\mathbf{a}_1\mathbf{p}|h_2|^2 + \mathbf{a}_2\mathbf{p}|h_2|^2 + \mathbf{a}_3\mathbf{p}|h_{12}|^2 + \sigma^2\right)$ and $g(\mathbf{p}) = -\log_2\left(\mathbf{a}_1\mathbf{p}|h_2|^2 + \sigma^2\right) - \log_2\left(\mathbf{a}_3\mathbf{p}|\tilde{h}_r|^2 + \sigma^2\right)$. Therefore, the Frank–Wolfe method of [114] can be adopted, which approximates $g(\mathbf{p})$ by its first-order Taylor series, and iteratively updates the first-order Taylor approximation along the specific direction that approaches the original function. The first-order Taylor approximation of $g(\mathbf{p})$ at the ith iteration is given by $g^{(n)}(\mathbf{p}) = -\log_2\left(\mathbf{a}_1\mathbf{p}^{(n)}|h_2|^2 + \sigma^2\right) - \log_2\left(\mathbf{a}_3\mathbf{p}^{(n)}|\tilde{h}_r|^2 + \sigma^2\right) - \frac{|h_2|^2\mathbf{a}_1(\mathbf{p}-\mathbf{p}^{(n)})}{\ln 2(\mathbf{a}_1\mathbf{p}^{(n)}|h_2|^2+\sigma^2)} - \frac{|h_r|^2\mathbf{a}_3(\mathbf{p}-\mathbf{p}^{(n)})}{\ln 2(\mathbf{a}_3\mathbf{p}^{(n)}|h_r|^2+\sigma^2)}$, where $\mathbf{p}^{(n)}$ is the value of $\mathbf{p}$ at the nth iteration. Now, the overall throughput is the difference between a convex function $f(\mathbf{p})$ and an affine function $g^{(n)}(\mathbf{p})$. We state Theorem 7.1 to solve the problem.

Theorem 7.1 *The reformulated problem, namely*

$$\frac{f(\mathbf{p}) - g^{(n)}(\mathbf{p})}{P_{\text{total}}(\mathbf{p})},$$

is jointly quasi-convex with respect to the vector variable $\mathbf{p}$ *in the feasible domain.*

Proof: Let us define the sublevel set of $\omega(\mathbf{p})$ as $S_\delta = \{\mathbf{p} \in \mathbf{\Theta} | \omega(\mathbf{p}) \le \delta\}$. Recall from [113] that $\omega(\mathbf{p})$ is jointly quasi-convex with respect to the variables in $\mathbf{p}$, if S_δ is convex for any real number δ. For $\delta \le 0$, we have no physical interpretation. By contrast, for $\delta \ge 0$, S_δ is equivalent to $S_\delta = \{[f(\mathbf{p}) - g^{(n)}(\mathbf{p})] - \delta P_{\text{total}}(\mathbf{p}) \le 0 | \mathbf{p} \in \mathbf{\Theta}\}$. According to our analysis above, $P_{\text{total}}(\mathbf{p})$ is affine with respect to the variables, while $[f(\mathbf{p}) - g^{(n)}(\mathbf{p})]$ is strictly jointly convex with respect to the variables. Therefore, the summation is strictly convex with respect to the variables, and $\omega(\mathbf{p})$ is quasi-convex with respect to the variables in $\mathbf{p} \in \mathbf{\Theta}$.

Theorem 7.1 confirms the optimality of the re-formulated objective function. For the fractional structured quasi-convex problem of Eq. (7.10), $\beta = \frac{f(\mathbf{p})-g^{(n)}(\mathbf{p})}{P_{\text{total}}(\mathbf{p})}$ can be associated with a subtract programming formulation of $f(\mathbf{p}) - g^{(n)}(\mathbf{p}) - \beta P_{\text{total}}(\mathbf{p})$ [114]. Therefore, with the aid of the equivalent subtract programming, the problem is reformulated as that of solving $f(\mathbf{p}) - g^{(n)}(\mathbf{p}) - \beta P_{\text{total}}(\mathbf{p})$ with a given β.

Now, we handle the constraints (C4) and (C5). Based on Eq. (7.7), (C4) is equivalent to $\mathbf{p}^{\mathrm{T}}\mathbf{A}\mathbf{p} + \mathbf{a}_4\mathbf{p} \le 0$, where $\mathbf{a}_4 = \left[0, (|h_2|^2 - |h_1|^2)\sigma^2, |h_{12}|^2\sigma^2\right]$ and

$$\mathbf{A} = \begin{bmatrix} 0 & 0 & |h_1|^2|h_{12}|^2/2 \\ 0 & 0 & |h_2|^2|h_r|^2/2 \\ |h_1|^2|h_{12}|^2/2 & |h_2|^2|h_r|^2/2 & |h_{12}|^2\sigma^2 \end{bmatrix}.$$

Since the matrix $\mathbf{A}$ is not a semi-positive definite matrix, its elements are not confined to a convex set. Therefore, we introduce the Schur complement [101] of $\mathbf{P} = \mathbf{p}\mathbf{p}^{\mathrm{T}}$ to relax (C4) into $(\hat{\mathrm{C}}4\mathrm{a}) : \mathrm{Tr}(\mathbf{AP}) + \mathbf{a}_4^{\mathrm{T}}\mathbf{p} \leq 0$ and constraint $(\hat{\mathrm{C}}4\mathrm{b}) : \begin{bmatrix} \mathbf{P} & \mathbf{p} \\ \mathbf{p}^{\mathrm{T}} & 1 \end{bmatrix} \succeq 0$. On the other hand, constraint (C5) is equivalent to $(\hat{\mathrm{C}}5) : \mathbf{a}_5\mathbf{p} + \Gamma_{2,\mathrm{req}}\sigma^2 \leq 0$, where we have $\mathbf{a}_5 = \left[\Gamma_{2,\mathrm{req}}\sigma^2, \ -|h_2|^2, \ -|h_{12}|^2\right]$. After a series of transformations, the problem becomes

$$\begin{aligned} \mathrm{P3} : & \underset{\mathbf{p}\in\{\boldsymbol{\Theta}\}}{\operatorname{argmin}} f(\mathbf{p}) - g^{(n)}(\mathbf{p}) - \beta P_{\mathrm{total}}(\mathbf{p}), \\ & \mathrm{s.t.} (\hat{\mathrm{C}}1) - (\hat{\mathrm{C}}3), (\hat{\mathrm{C}}4\mathrm{a}), (\hat{\mathrm{C}}4\mathrm{b}), \text{and } (\hat{\mathrm{C}}5). \end{aligned} \tag{7.11}$$

The problem P3 in Eq. (7.11), now, is a standard semi-definite programming (SDP) problem with a convex set, which can be readily solved by the CVX package of MATLAB. Finally, a so-called "EE oriented FD cooperative NOMA" algorithm is introduced for optimizing $\mathbf{p}$, which solves the SDP problem P3 of Eq. (7.11) in the inner layer and updates β in the outer layer. To tighten the first-order Taylor approximation $g^{(n)}(\mathbf{p})$, we update the value of $\mathbf{p}^{(n)}$ in the specific direction that approaches the original function. Since $f(\mathbf{p}) - g^{(n)}(\mathbf{p}) - \beta P_{\mathrm{total}}(\mathbf{p})$ serves as the upper bound of the original problem $f(\mathbf{p}) - g(\mathbf{p}) - \beta P_{\mathrm{total}}(\mathbf{p})$, the value of the upper bound is reduced iteratively until convergence is reached, as suggested by Line 4 of Algorithm 7.1. Furthermore, the value of β is squeezed by the bisection method in the outer layer according to the accuracy factor ϵ, and the optimal $\beta^\star$ can be found after a few updates. Hence, the tightness and convergence of the algorithm is confirmed. The associated procedure is summarized in Algorithm 7.1.

Algorithm 7.1 EE Oriented FD Cooperative NOMA Algorithm

Input: Left/right bounds β_{l} and β_{r}, channel condition, i.e., $h_1, h_2, h_{12}, \widetilde{h}_{\mathrm{r}}$, and power consumption parameters, i.e., $\eta, p_{\mathrm{c,t}}, p_{\mathrm{r,t}}, p_{\mathrm{SI}}, p_{\mathrm{fix}}$.

Output: Optimal transmission power vector to $\mathbf{p}^*$.

1: Set the accuracy factor to $\epsilon > 0$, and assume that $\mathcal{F}(\beta)$ is the optimal value of $f(\mathbf{p}) + g^{(n)}(\mathbf{p}) - \beta P_{\mathrm{total}}(\mathbf{p})$. Initialize the left bound β_l and the right bound β_r for ensuring that $\mathcal{F}(\beta_{\mathrm{l}}) \cdot \mathcal{F}(\beta_{\mathrm{r}}) < 0$.
2: **while** $\beta_{\mathrm{r}} - \beta_{\mathrm{l}} > \epsilon$ **do**
3:     $\beta = \frac{\beta_{\mathrm{r}} + \beta_{\mathrm{l}}}{2}$.
4:     Solve the problem P3 using the Frank-Wolfe method until convergence.
5:     **if** $\mathcal{F}(\beta_{\mathrm{l}}) \cdot \mathcal{F}(\beta) < 0$ **then**
6:         $\beta_{\mathrm{r}} = \beta$.
7:     **else**
8:         $\beta_{\mathrm{l}} = \beta$.
9:     **end if**
10: **end while**

Let us now consider the complexity of the algorithm. Assume that $\beta_1 = \frac{\beta_r+\beta_l}{2}$ is the midpoint of the initial interval, and β_n is the midpoint of the interval in the nth step. Then the difference between β_n and β^* is bounded by $|\beta_n - \beta^*| \leq \frac{\beta_r-\beta_l}{2^n}$. Given a tolerance factor ϵ, the required number of iterations is given by $n \leq \log_2 \frac{\beta_r-\beta_l}{\epsilon}$. In the algorithm, the left bound β_l can be set to 0. Then, the value of $\mathcal{F}(\beta_l)$ is definitely positive. Furthermore, a sufficient large value of β_r can be chosen as the right bound for making the value of $\mathcal{F}(\beta_r)$ negative. Therefore, the function $\mathcal{F}(\cdot)$ has opposite signs at the two bounds and thus the classic bisection method readily leads to convergence. In the inner layer, the CVX solver invokes an interior-point method to solve the SDP problem, which belongs to the class of path-following methods and leads to a rapid convergence. Upon denoting the complexity of the inner layer by ξ, the complexity order of the algorithm becomes $\mathcal{O}(\log_2(\frac{\beta_r-\beta_l}{\epsilon}) \cdot \xi)$.

Remark 7.1 Generally speaking, a throughput maximization problem aims for achieving a high throughput regardless of the power consumption. Hence, the BS only has to satisfy the WU's SINR threshold, while allocating all the remaining power to the SU. Then, the SU utilizes a high transmission power to cooperate with the WU. By contrast, a power minimization problem aims for reducing the total power consumption, and thus merely satisfies the WU's QoS constraint by a minimum transmission power. As a better choice, EE maximization improves the ratio between the throughput and power consumption, striking a trade-off between the throughput maximization and power minimization problems. As suggested by Theorem 7.1, $\omega(\mathbf{p})$ is quasi-convex (thus the re-formulated EE is quasi-concave) with respect to the variables in $\{\boldsymbol{\Theta}\}$. For a quasi-concave function, utilizing all transmission power may not lead to the most optimal EE solution, because the power consumption in the denominator is also boosted, hence the EE is decreased. For illustration, let us consider a simple EE maximization problem in Fig. 7.2, where the EE curve is a quasi-concave function of transmission power. Neither the power minimization (point (a)) nor the throughput maximization solution (point (c)) achieves the optimal EE.

Remark 7.2 As suggested by Eqs. (7.4), (7.6), and (7.7), the BS-SU, BS-WU, and SU-WU distances also affect the SIC failure region. Given a shorter BS-SU distance, the value of Γ_{12} is increased due to the reduced path loss (PL), and thus a higher cooperative power may be assigned to the SU hence guaranteeing $\Gamma_{12} \geq \Gamma_2$. A similar trend can be observed for a shorter BS-WU distance. By contrast, a longer SU-WU distance makes $\Gamma_{12} \geq \Gamma_2$ more likely. However, this also makes the cooperation between the SU-WU less energy efficient, since additional power is consumed at the SU without a reasonable throughput improvement. As expected, when the SU-WU distance tends to infinity, the cooperative power assigned to the SU tends to 0, as in non-cooperative NOMA systems.

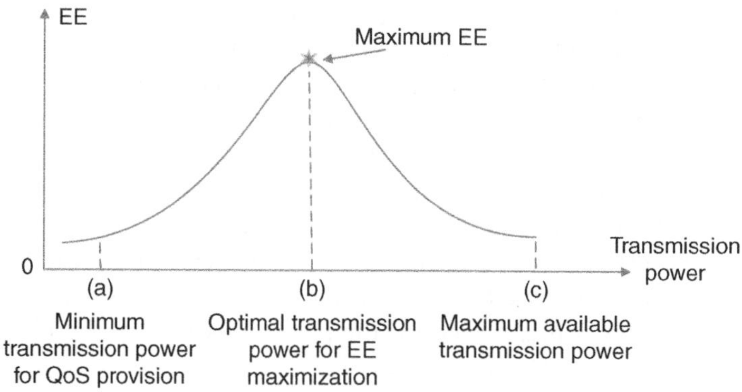

Figure 7.2 The stylized relationship of EE maximization, power minimization, and throughput maximization problems.

Remark 7.3 Given a lower drain efficiency, the system's EE is reduced owing to the boosted PA power consumption. Furthermore, the power allocation of both the BS and of the SU becomes more conservative due to the low drain efficiency, since the throughput improvement attained by increasing the transmission power may not be in line with the increased power consumption.

7.4 Simulation Results

Now, we discuss the numerical results to verify our analysis using the parameters given in Table 7.1. The PL model of $\text{PL}(d) = 145.4 + 37.5\log_{10}(d/1000)$ [114] and log normal shadowing model are adopted, which are featured in the 3GPP long-term evolution (LTE) standards operating at 2 GHz. The small-scale fading

Table 7.1 Simulation Setup.

Bandwidth	1.25 MHz
AWGN power spectral density	−174 dBm/Hz
Drain efficiency of PA η	35%
$p_{\text{c,r}}$ and $p_{\text{c,t}}$	100 mW
p_{SI} and p_{fix}	50 and 500 mW
$p_{\text{BS,max}}$ and $p_{3,\text{max}}$	100 and 20 mW
SI cancellation amount	80 dB
Distance between the BS and the WU	200 m

is modeled by Rayleigh fading except for the SI channel at the SU, which is modeled as Rician fading and having a Rician factor of 5 dB [110]. A pair of typical cooperative NOMA systems are selected for performance comparison. (a) In HD cooperative NOMA systems [115], the BS transmits its downlink signals to the two users (with transmission power $p_1 = \alpha p_{\text{BS,max}}, p_2 = (1-\alpha)p_{\text{BS,max}}$, and $\alpha = 0.2$) during the first half of the time slot, while the SU helps the WU by assigning its full transmission power $p_3 = p_{3,\text{max}}$ in the second half of the time slot. (b) In the pre-fixed FD cooperative NOMA systems [106], the SU acts as an FD relay node for simultaneously helping the WU by assigning its full transmission power $p_3 = p_{3,\text{max}}$, whereas the power allocation at the BS is pre-fixed as $p_1 = \alpha p_{\text{BS,max}}, p_2 = (1-\alpha)p_{\text{BS,max}}$, and $\alpha = 0.2$.

Figure 7.3(a) and (b) show the impact of the normalized distance between the BS and the SU on the EE and the spectrum efficiency (SE), respectively. As can be seen, the EE-oriented FD cooperative NOMA algorithm outperforms the others in terms of its EE, whilst exhibiting a higher robust against the SU's location as well. This is because the benchmarking algorithms consume all the available transmission power, which degrades their EE. A further degrading factor is constituted by their low throughput. In Fig. 7.3(b), the EE-oriented FD cooperative NOMA algorithm shows the highest throughput among the three algorithms. This is because the EE-oriented FD cooperative NOMA algorithm supports the effective

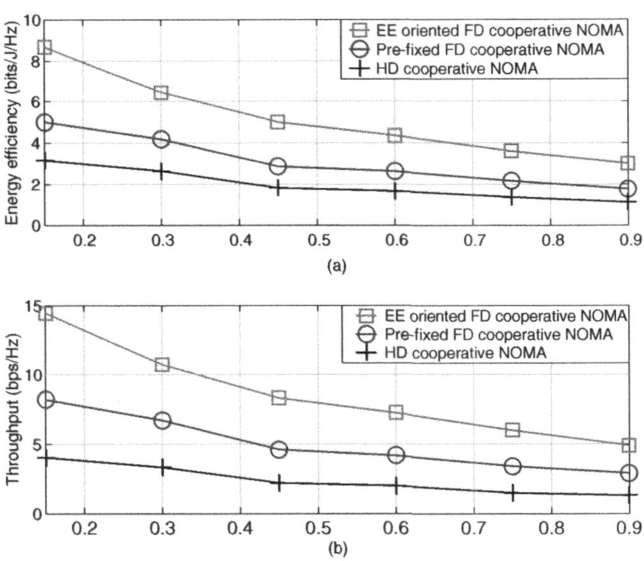

Figure 7.3 Impact of the normalized distance between the BS and the SU on the value of EE and throughput, with $p_{3,\text{max}} = 20 \times 10^{-3}$ mW. (a) normalized distance between the BS and the SU and (b) normalized distance between the BS and the SU.

cooperation of the SU and of the WU, hence improving the WU's throughput. More importantly, no additional time slots are required for the cooperation phase, and thus the EE-oriented FD cooperative NOMA algorithm significantly outperforms the HD cooperative NOMA. As for the pre-fixed FD cooperative NOMA, the power allocation is unaware of the relative distance among the communication nodes. Furthermore, the SU invokes its full transmission power to cooperate with the WU, which may result in the "SINR gap reversal" and may lead to an unsuccessful SIC operation at the SU. As a result, the throughput of the pre-fixed FD cooperative NOMA is also inferior to the EE-oriented FD cooperative NOMA algorithm. Finally, the throughput of all the algorithms degrades for a longer distance between the BS and the SU due to the high PL.

Figure 7.4 shows the impact of the constraint $p_{3,\max}$ on the SIC's failure probability, which is defined as the ratio of the times of unsuccessful SIC operation at the SU to the times of simulations. As we can see, the EE-oriented FD cooperative NOMA algorithm is shown to guarantee that the SU can successfully decode the WU's signal by SIC. This is because an additional constraint is imposed on the power control at the SU by the EE-oriented FD cooperative NOMA algorithm. By comparison, the SIC failure probability of the pre-fixed FD cooperative NOMA

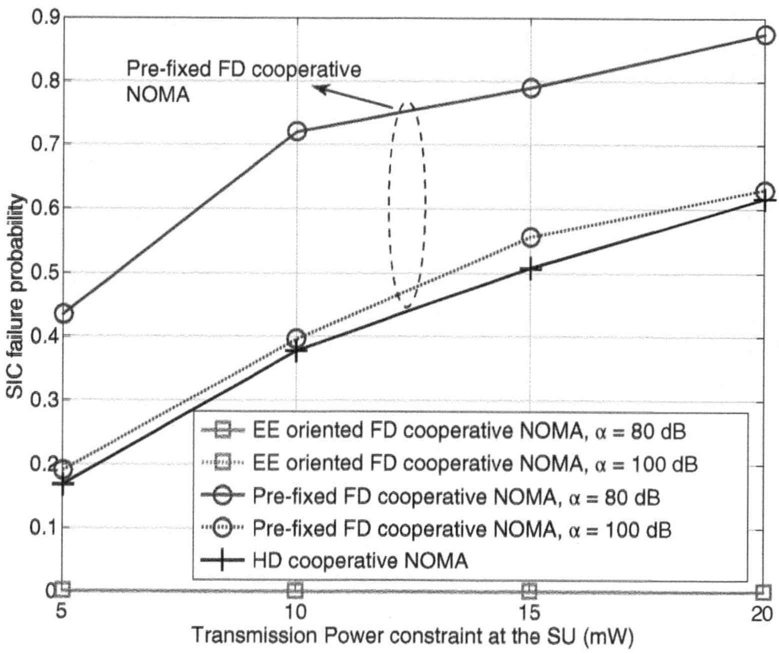

Figure 7.4 Impact of the maximum constraint $p_{3,\max}$ at the SU on the SIC's failure probability, where the SU is in the middle of the BS and the WU.

systems increases rapidly with the transmission power constraint $p_{3,\max}$. This is because for a higher transmission power used at the SU, the WU's SINR Γ_2 increases, while the WU's SINR Γ_{12} observed at the SU decreases due to the higher level of residual SI. As a result, a higher value of $p_{3,\max}$ leads to a higher SIC failure probability at the SU. In case of poor SI cancellation, the SIC failure probability is substantially increased by the strong residual SI. As for the HD cooperative NOMA system, its SIC failure probability also increases upon imposing a higher transmission power constraint due to the "SINR gap reversal" problem. Since no residual SI is imposed by the HD operation, its SIC failure probability is lower than that of the FD operation [106]. However, its SIC failure probability still remains significantly higher than that of the EE-oriented FD cooperative NOMA algorithm, because the HD cooperative NOMA neglects that Γ_2 may be higher than Γ_{12} owing to the cooperation between the SU and WU. However, its SIC failure probability is not affected by the SI cancellation factor α in the presence of HD operation.

7.5 Conclusion and Remarks

In this chapter, we introduced an EE-oriented algorithm for FD cooperative NOMA systems, where the transmission power of both the BS and of the SU is adaptively allocated, rather than being pre-fixed. More importantly, the "SINR gap reversal" issue of cooperative NOMA systems is solved by adaptively confining the transmission power at the SU.

Note that the intrinsic intra-cluster multi-device interference, also deemed as an underlying resource, has not been fully exploited when designing NOMA schemes. Considering the high device density in IoT systems, if the rich multi-device interference can be utilized rather than being mitigated, the system performance can be significantly enhanced. Therefore, in the next chapter, we will outline the design principles of NOMA in partially-coordinated beamforming and fully-coordinated joint transmission.

8

Signal Design for Multicluster Coordination

- The imminent Internet of Things (IoT) aims to provide massive device connectivity and support ever-increasing data demands, making today's environment on the edge of a new era of innovations and changes. In multi-cluster IoT, the devices may suffer severe inter-cluster interference due to the intensive frequency reuse among adjacent access points (APs), thus deteriorating their quality-of-service (QoS). To address this issue, conventional multi-cluster coordination in IoT provides orthogonal code-, frequency-, time-, or spatial-domain multiple access for interference management, which yet results in a waste of resources, especially in the context of explosively increased number of devices. In this chapter, we review the recent advances in energy-efficient solutions for multi-cluster coordination relying on non-orthogonal multiple access (NOMA) and constructive interference (CI) exploitation techniques. Moreover, their applications for handling interference management are elaborated at different levels of multi-cluster coordination. Departing from traditional orthogonal multiple access (OMA), this line of research provides a new dimension for inter- and intra-cluster multi-device interference management in IoT.

8.1 Multi-cluster Coordination in IoT

IoT networks aim to provide heterogeneous services for massive devices, where the use cases can be roughly classified into massive machine-type communications (MMTC), and ultra-reliable and low latency communications (URLLC) [116]. Specifically, MMTC is featured by massive connectivity, low power consumption, and high energy efficiency, while URLLC is mainly designed to provide high reliability and low latency for serving mission-critical applications [117].

IoT Signal Detection, First Edition. Rui Han, Jingjing Wang, Lin Bai, and Jianwei Liu.

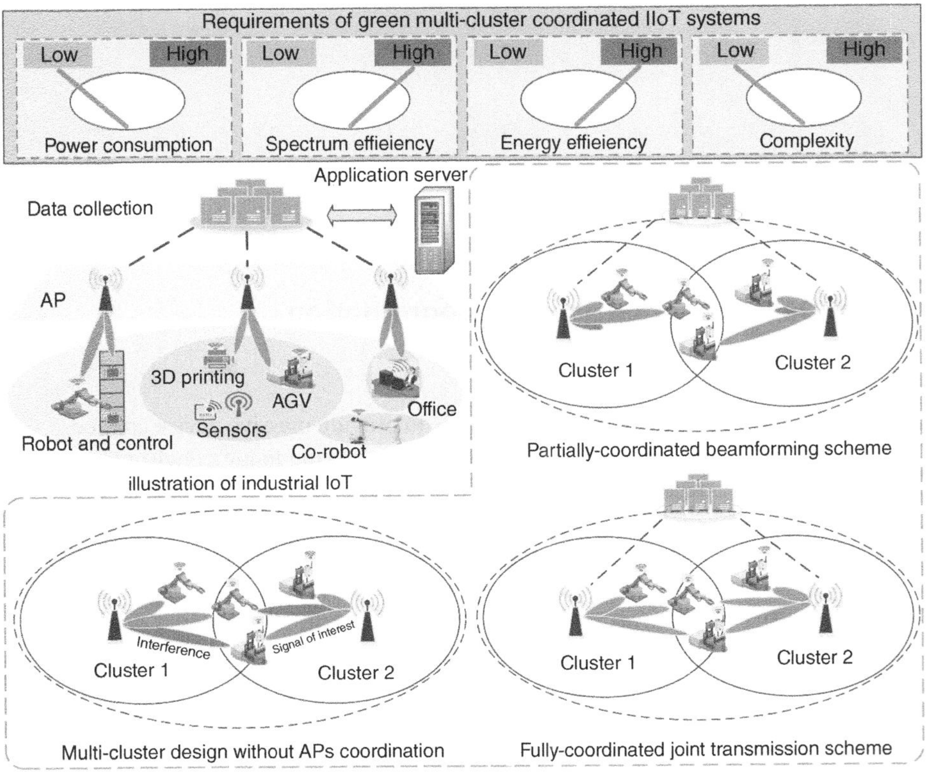

Figure 8.1 Low energy consumption, high energy efficiency, and low complexity techniques are preferable to enable an energy-efficient multi-cluster coordinated IIoT system.

Evidently, wireless communications in IoT cause severe access congestion due to insufficient frequency, time, power, antennas, and other resources. One of the efficient techniques is to cluster massive devices into a number of small groups, with a part of resources being shared and reused in different clusters, referred to as multi-cluster IoT [118]. A typical cluster-based industrial IoT (IIoT) is depicted in Fig. 8.1. Each cluster is equipped with one or more APs to provide high-quality wireless transmissions. A number of APs are connected by a high-speed optical fiber to data collection module and application server for resource coordination, authentication, and interference management. Therefore, the wireless interface management between APs and devices is essential in IoT, from the perspectives of radio spectrum allocation and interference management [119, 120].

According to the coordination level among APs, the multi-cluster coordination techniques in IoT can be mainly categorized into two modes: partially-coordinated beamforming (PBF) and fully-coordinated joint transmission (FJT) [117]. For the PBF mode, channel state information (CSI) is shared by a number of APs to design cooperative beamforming for suppressing inter-cluster interference [121, 122]. For the FJT mode, in addition to CSI, the intended transmission data are shared [123]. Since both CSI and the corresponding transmission data are available at all the coordinated APs, the FJT mode allows the coordinated APs to transmit the same data to the corresponding devices. Therefore, the FJT mode provides an improved degree-of-freedom (DoF) in terms of spatial diversity gain and achieves superior system performance than the PBF mode.

Generally, since only CSI is shared among APs, the beamforming design of the PBF system makes a trade-off between inter-cluster interference suppression and maximizing the devices' signal-to-interference-plus-noise ratio (SINR) within the cluster of interest, with a moderate level of coordination overhead. On the other hand, the FJT scheme can be regarded as a virtual multiple-input multiple-output (MIMO) system where, all AP-device links are utilized to convey data, at the expense of higher coordination overhead due to the CSI and data sharing mechanism among the APs [118].

These multi-cluster coordination techniques have distinct advantages and disadvantages, whereas their implementation in IoT is still restrictive. It is because:

- Due to the massive connections of devices, intra-cluster (multi-device) interference within each cluster becomes a critical limitation for ensuring QoS target of devices. In terms of multi-device interference management, by the existing coordination systems, wireless resources, such as time and frequency can only be exclusively assigned to one device on the basis of OMA [124, 125]. Nevertheless, the limitations of the orthogonality-based methods are increasingly extrusive. Since IoT is typically loaded with massive equipments, the conventional orthogonal approaches inevitably require extra time and frequency resources for providing strict orthogonality, otherwise only part of devices can be served at the cost of high latency and poor fairness.
- Apart from the orthogonal time/frequency access, another approach is based on the interference alignment (IA) technique, which provides spatial orthogonality for multi-device access. Nevertheless, it requires large-scale antennas to achieve strict spatial orthogonality by carefully designing the beamforming vector to cancel multi-device interference, which is cost/power consuming and thus may not be suitable for the small-sized APs.
- In IoT communications, there is an urgent demand for providing high system throughput while limiting energy consumption. Hence, energy efficiency (EE) defined as the ratio of throughput to total power consumption has attracted much attention from vendors and researchers, where the conventional throughput maximization-oriented designs in cellular networks become incongruous.

Since NOMA is able to multiplex a large number of devices onto the same frequency resource, it is particularly suitable for the MMTC application due to its enhanced multi-device access capability. Its applications in IoT have been researched in terms of the NOMA-based parallel spaceborne antenna array calibrations and impulsive noise mitigation. Nevertheless, they merely applied NOMA for non-coordination IoT, where the system performance may be significantly impaired by the intra-cluster interference. Furthermore, note that the intrinsic intra-cluster multi-device interference, also deemed as an underlying resource, has not been fully exploited when designing non-orthogonal transmission techniques. Considering the high device density in IoT, if the rich multi-device interference can be utilized rather than being mitigated, the system performance can be significantly enhanced. This inspires the interest in treating the ubiquitous interference in IoT as a green source with the aid of the spatially non-orthogonal design, namely constructive interference (CI).

8.2 Multi-cluster Coordination with NOMA

As mentioned above, the conventional multi-cluster coordination methods (such as the PBF and FJT) provide orthogonal code-, time-, frequency-, or spatial-domain multi-device access for interference management, which strictly limit the number of devices being served. To alleviate this, it is imperative to collaborate NOMA with multi-cluster coordination to provide flexible multi-device access, which also enables an improved level of DoFs in resource allocation and interference management. Together with the low power and high EE demands of IoT communications, the green multi-cluster coordinated NOMA IoT system has become an essential and practical issue. In this section, we first outline the design principles of NOMA in PBF and FJT. Then, the device association (DA), successive interference cancellation (SIC) design, device grouping, and power control policies tailored for IoT are detailed.

8.2.1 Multi-cluster Coordination NOMA Design

Based on the shared information among the coordinated APs, the multi-cluster coordinated NOMA techniques can be classified into partially-coordinated beamforming NOMA (PBF-NOMA) and fully-coordinated joint transmission NOMA (FJT-NOMA).

1) **PBF-NOMA**: In the PBF-NOMA, the devices are served by their associated local APs, and the devices within each cluster are scheduled into NOMA groups

[126, 127]. Based on the shared devices' CSI among the APs, inter-cluster interference is suppressed by the PBF, while intra-cluster interference is handled by SIC design. Compared to the conventional PBF-OMA designs, the PBF-NOMA scheme can simultaneously support more device connections and provide a high level of DoFs in multi-device interference management.

2) **FJT-NOMA**: In the FJT-NOMA, the devices can be jointly served by the coordinated APs, where a global manner NOMA is achieved among all the coordinated clusters. Hence, this higher level of coordination system is more device-centric, as the devices can be scheduled into the same groups even if they locate into different APs' coverage [128, 129]. In addition, compared to the PBF-NOMA scheme where inter-cluster interference is extensively suppressed, the FJT-NOMA takes advantage of the inter-cluster interference channel for data transmission and benefits from the spatial diversity gain from the distributed APs [128, 129].

For the FJT mode, each AP shares CSI and the transmitted data to other coordinated APs. With K devices in each AP's coverage and N coordinated APs, the total overhead of sharing CSI and data is given $O\left(N(N-1)K\left(\chi_C+\chi_S\right)\right)$, where χ_C denotes the required bits for describing one device's CSI while χ_S denotes the required bits for exchanging each device's data symbols. Generally, only a few bits are enough to describe a multiple-input single-output (MISO) channel for each user [130]. For the overhead on sharing data symbols, it is decided by the downlink frame structure of the communication protocol. Hence, the overhead on sharing data symbols is reasonable, considering the fact that the APs have been connected to a centralized cloud-manager through high-speed optical fibers.

8.2.2 Multi-cluster Coordinated NOMA Resource Allocation

1) **Device association**: DA plays a significant role in load balancing and performance enhancement. The existing DA designs can be generally classified into distance-based and channel gain-based methods. In particular, the distance-based method is sensitive to channel fading and shadowing, while the performance of the channel gain-based method may be significantly impaired by the inter-cluster interference. As a result, multiple criteria, such as distance, channel gain, and inter-cluster interference, and so on, should be jointly considered to make a comprehensive decision for DA. Recently, artificial intelligence (AI) techniques have been proposed to strike an attractive balance among different criteria, enhancing the robustness against the effects of channel fading, shadowing, and inter-cluster interference [131].
2) **SIC design**: SIC is employed at the receive side for superposed signal decoding and demodulating. There are two fundamental rules: (Rule 1) the optimal

SIC decoding order of devices is based on the increasing order of their channel gains from the APs [127], and (Rule 2) the rate of the device with a higher decoding order (e.g., the device i) to detect the signal of the device j should be no lower than the targeted rate of the device j, so that the signal of the device j can be successfully decoded and removed at the device i. For the single-cluster NOMA IoT, Rule 1 is equivalent to Rule 2. Nevertheless, in the multi-cluster NOMA, due to the incurred inter-cluster interference, the rate of the device i to decode the device j may be lower than the targeted rate of the device j, which results in the failure of SIC process. Therefore, to successfully perform SIC, dedicated SIC design considering the effect of the inter-cluster interference should be developed to guarantee the validity of Rule 2 [128, 129].

3) **Device grouping and power control**: For the PBF-NOMA, the devices within each cluster can be scheduled into a number of NOMA groups, and power allocation policies are carefully designed by the coordinated APs to suppress the intra-cluster inter-group interference. On the other hand, for the FJT-NOMA scheme, both the CSI and transmission data are available at coordinated APs. As a result, the devices located in the different clusters could be scheduled into the same NOMA group, thereby enabling a higher DoF in device grouping. In addition, since the inter-cluster interference channels from the coordinated APs to devices can be utilized to carry useful information, only the intra-cluster inter-group interference needs to be suppressed, which results in lower power consumption and higher EE performance.

By applying the multi-criterion-based DA, device grouping, and power allocation schemes, the multi-cluster coordinated IoT can provide an energy-efficient transmission [131]. Figure 8.2(a) and (b) shows the power consumption and EE with different coordination designs. It can be observed that, with stringent transmission rate requirements, the multi-cluster FJT-NOMA scheme consumes lower transmission power and achieves higher EE performance over the non-coordinated multi-cluster NOMA design, i.e., EE enhancement of more than 27.4% and transmission power reduction of around 34.6 dBm. In addition, the multi-criterion-based DA scheme provides 16.8% EE enhancement, and 31.1 dBm transmission power reduction than the single-criterion-based methods, enabling a greener multi-cluster coordinated NOMA IoT system. Note that though SIC incurs additional power at the receiver sides, typically ranging from −3 to 20 dBm, this only contributes a tiny proportion to the total power consumption. Hence, the amount of power saving achieved by the NOMA-based multi-cluster design overwhelms the additional power consumption incurred by the SIC operation.

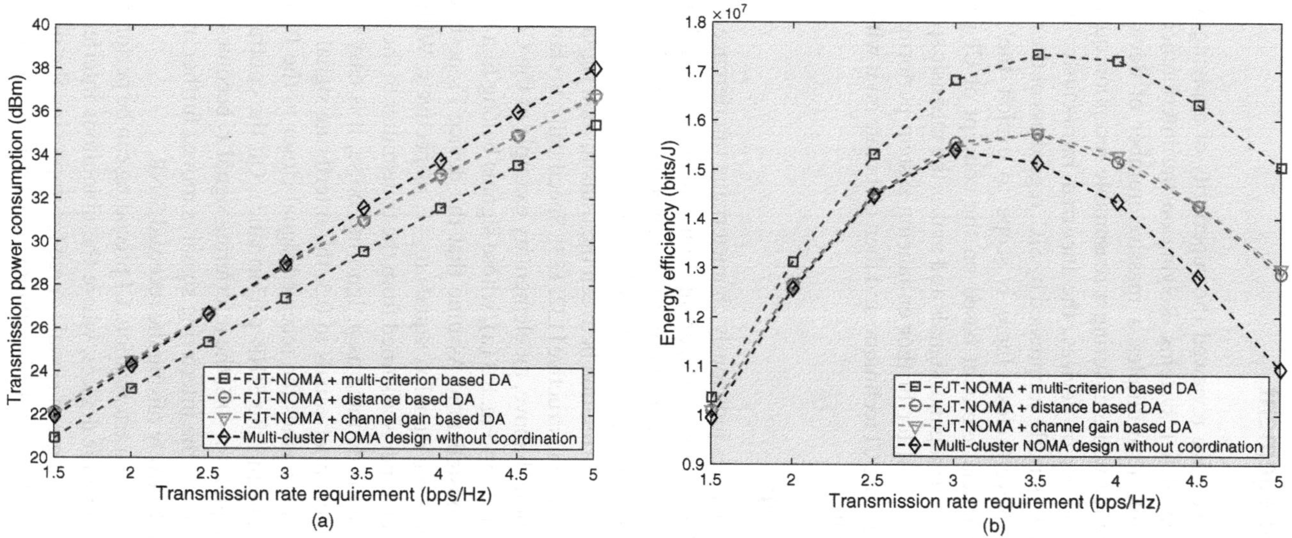

Figure 8.2 Transmission power consumption and EE performance with 5 MHz bandwidth. (a) transmission power consumption performance and (b) EE performance with different coordination designs.

8.3 CI-aided Multi-cluster Coordination with Interference Management

As discussed above, we have reviewed the superiority of applying NOMA into multi-cluster coordinated IoT systems. Relying on the interference-mitigation-based design, it is able to multiplex a massive number of devices onto the same frequency resource while ensuring a reasonable reception quality. Due to the non-orthogonal access of devices, the incurred intra-cluster multi-device interference needs to be strictly suppressed by SIC, regardless of the multi-cluster coordination levels. Nevertheless, there is scope to exploit the multi-device interference as a beneficial element based on the concept of CI precoding. It transmits spatially non-orthogonal signals and makes the intrinsic spatial leakage constructive [130, 132, 133] for further enhancing system performance. In this section, we first review the CI technique and then elaborate on it with the cluster coordination techniques.

8.3.1 CI Signal Design

The CI exploitation involves judiciously characterizing interference by considering the signal constellation size. Departing from the conventional interference suppressing techniques which constrain the received signals within a proximity area around the modulated signal constellation point, the CI scheme can exploit, rather than suppress, the correlation between the transmission signal to make the multi-device interference constructive [132, 133]. Specifically, to clarify the fundamental concept mentioned above, an elementary example of the CI exploitation with two devices is given in Fig. 8.3 (a), whose signals belong to a binary phase shift keying (BPSK) constellation. Assume that the device 1's desired signal x_1 equals to 1 and interfering signal x_2 equals to −1, respectively. Without loss of generality, a lossless channel is assumed from the transmitter to the device 1, and the interfering channel is denoted as ρ. Ignoring noise, it is easy to verify that when the interfering channel equals to 0.5, the interfering signal x_2 is destructive to the device 1 as it pushes the received signal closer to the BPSK decision threshold, thereby reducing the SINR performance. On the contrary, when the interfering channel equals to −0.5, the interfering signal x_2 becomes constructive to the desired signal x_1, as the interfering signal is moved further away from the decision threshold, effectively raising the received SINR.

On the basis of the CI characterization, CI precoding can be performed to make interference constructive for devices, where the ubiquitous interference is judiciously transformed into a green signal source for improving the devices' reception performance. In other words, by employing CI, lower transmission power is required at the devices to achieve a targeted performance since interference contributes constructively rather than being suppressed. The above fundamental

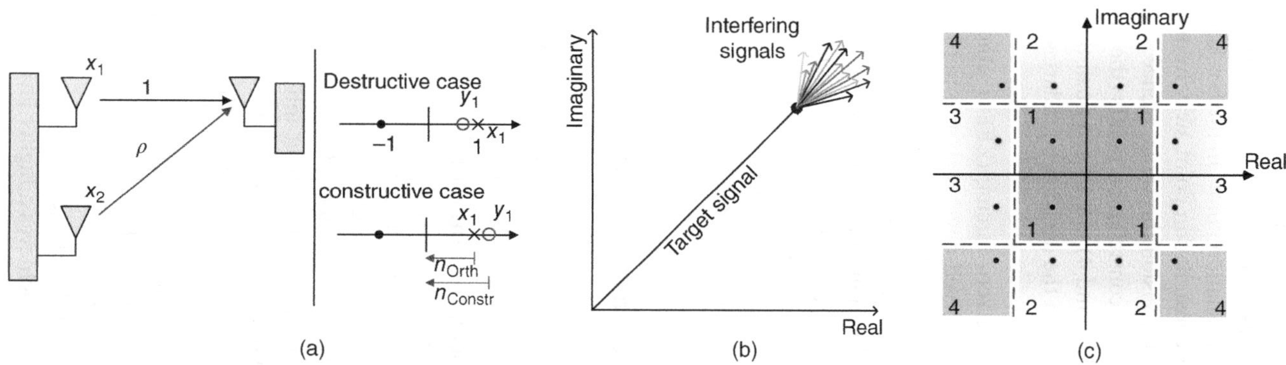

Figure 8.3 (a) An elementary example of CI exploitation with BPSK constellation; (b) A quadrature phase shift keying (QPSK) constellation example with CI precoding exploits interference as a beneficial element; (c) A schematic representation of 16-QAM constellation points.

example can be extended to general M-order of phase shift keying (PSK) (as shown in Fig. 8.3 (b)) and quadrature amplitude modulation (QAM) (as shown in Fig. 8.3 (c)) [133]. Take 16-QAM for illustration, it can be seen that for the inner constellation points (the group of constellation points in the box labelled 1), interference that shifts the received inner constellation point away from one decision threshold pushes it closer to another decision threshold. Hence, the concept of CI does not hold. However, for the outer constellation points (the group of constellation points in the boxes labeled 2, 3, and 4), the CI can push the received signals fall in the detection region away from the decision boundaries to enhance the signal reception performance.

Moreover, as for the existing zero forcing (ZF) or other optimization-based precoders, the number of the served devices should not be larger than that of the transmit-antennas. By contrast, the CI locates the desired signals into constructive regions (where the received signals have increased the distance to the detection threshold of demodulation), rather than strictly locating the signals in the proximity region around the constellation point. With a higher design DoF at the transmitter-side, the number of the served devices can be larger than that of the transmit-antennas [130, 132], while achieving an enhanced reception performance. Hence, the CI is particularly suitable for massive downlink-connectivity in MMTC scenario.

8.3.2 CI Design for Multi-cluster Coordination

The key challenge of integrating CI and multi-cluster coordination lies in the inter-cluster interference presented from the adjacent APs and the multi-device interference within each cluster, which need to be separately managed.

1) **Partially-coordinated beamforming CI (PBF-CI)**: In the PBF-CI scheme, since CSI is available at the coordinated APs, the intra-cluster multi-device interference can be predicted and characterized before transmission, which then can be treated as a beneficial element at the receiver side, enabled by the CI precoding. However, the intended data is merely transmitted by the local APs, which makes it difficult to exploit inter-cluster interference as a constructive element. To this end, the philosophy of the PBF-CI is to exploit intra-cluster multi-device interference while carefully mitigating inter-cluster interference [130].
2) **Fully-coordinated joint transmission CI (FJT-CI)**: In the FJT-CI, both CSI and the intended transmission data of the devices are shared among the coordinated APs. As a result, not only the inherent intra-cluster multi-device interference but also the inter-cluster interference can be exploited as constructive elements, which achieves a higher level of DoFs in interference management and significantly enhances performance, at the cost of higher coordination overhead over the PBF-CI scheme.

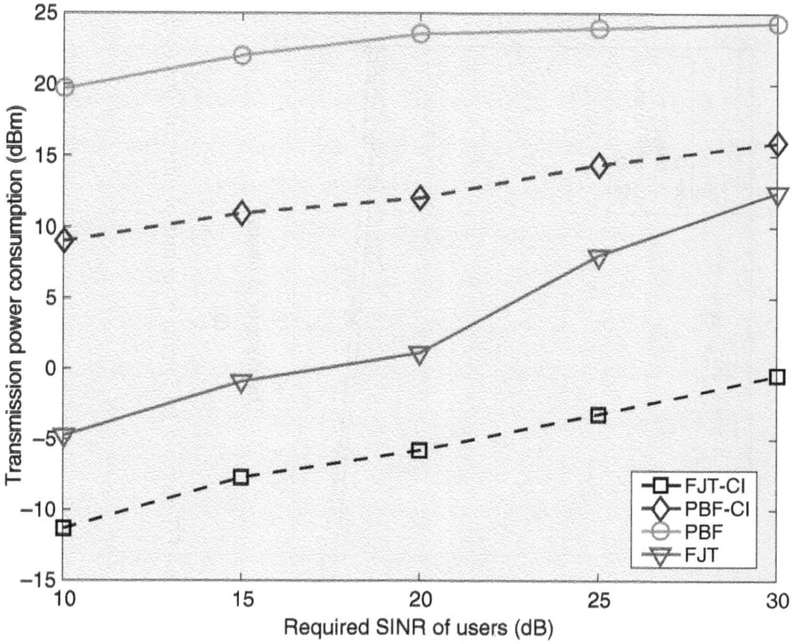

Figure 8.4 The performance gain of the FJT-CI and PBF-CI designs over the conventional coordination techniques.

The impact of the devices' SINR requirements on the total transmission power is shown in Fig. 8.4. As can be seen, the FJT-CI scheme consumes the lowest power, providing 12.2 dBm power saving over the FJT, 15.8 dBm over the PBF-CI, and 24.2 dBm over the PBF scheme, respectively. The reason lies in that the FJT-CI scheme utilizes both inter-cluster and intra-cluster multi-device interference as constructive elements, which leads to much lower transmission power for achieving a targeted SINR. In addition, the FJT-CI can also benefit from the spatial diversity gain by taking advantage of the inter-cluster interference channel, thanks to the distributed coordinated APs. For the PBF-CI scheme, as the intended transmission data is not shared among the APs to reduce coordination overhead, only intra-cluster multi-device interference can be made constructive while inter-cluster interference should be carefully mitigated by joint precoding design. It is clear from the above results that the employment of the CI exploitation in IoT has the potential of orders-of-magnitude reduction in power consumption.

The CI design also has a high potential in the URLLC scenario. The symbol error rate (SER) and execution time performance are demonstrated in Fig. 8.5. It can be concluded that the CI endorses the lowest SER at moderate/high SINR

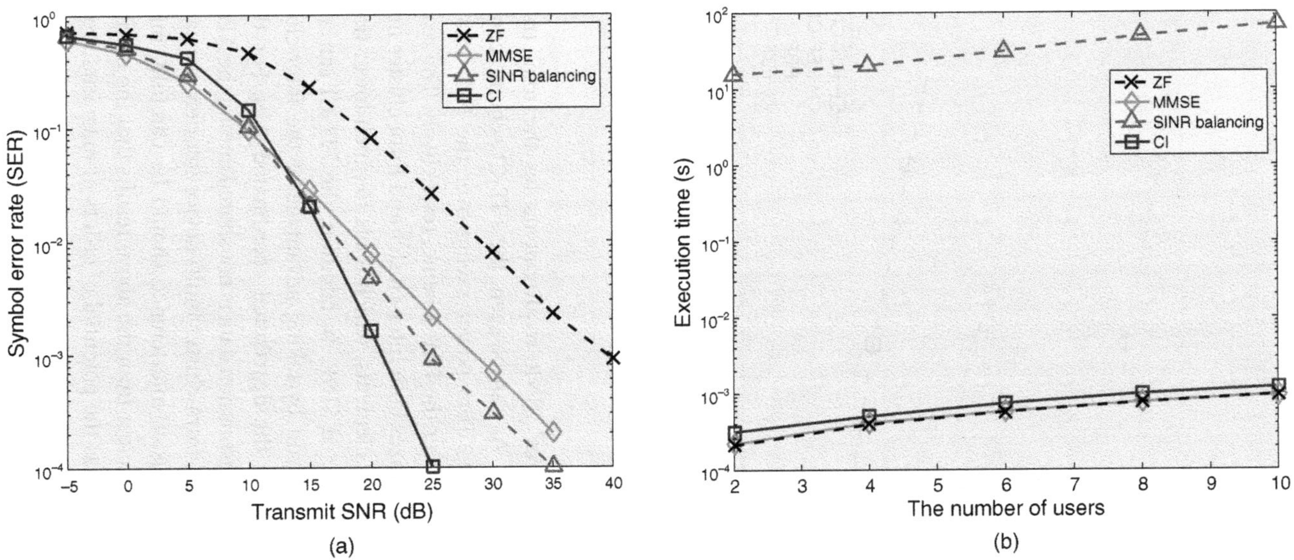

Figure 8.5 The symbol error rate and execution time of different precoders in the FJT scenario. (a) Symbol error rate of different precoders and (b) execution time of different precoders.

regimes, and thus outperforms the ZF, minimum mean square error (MMSE), and SINR balancing precoders in terms of reliability performance. In addition, the CI design requires the same execution time as the ZF and MMSE precoders, which are known as the most practical precoders due to their low complexities. As a result, the CI design imposes no additional latency on the signal transmission, and is readily compatible with existing URLLC-dedicated techniques.

8.4 Future Works

1) **Hybrid NOMA and CI transmission for a multi-cluster coordination IoT**: NOMA is able to provide massive device access with a limited number of antennas, while its performance may be significantly reduced if the devices are with high-channel correlation. To this end, hybrid NOMA and CI transmission can be exploited for multi-cluster coordination, according to the devices' distinct traffic demands and channel conditions. For example, devices with an instant access request and high channel disparity can be served by NOMA, while devices having channel correlations and requiring high throughput can be served by CI. Hence, how to adaptively provide the synthetic NOMA and CI transmission while guaranteeing devices' QoS requirements is still an open issue.
2) **Multi-cluster coordination IoT with fog network computing**: Fog computing is envisioned as a crucial technology in IoT to provide low latency and local processing, by offloading the computing and functionalities storing from the centralized cloud center to the edge-fog devices. Nevertheless, how to integrate fog computing-based radio access networks (F-RAN) with multi-cluster coordination is still challenging. It requires a new topology design, such as the design and allocation of fog nodes across multiple clusters. Since the fog devices in F-RAN are supplemented with additional computation capacity, how to jointly leverage the computation capacity of fog nodes with low overhead is yet to be explored.
3) **Multi-cluster coordination design for devices with limited statistical or unavailable CSI**: The existing coordination techniques generally rely on the availability of CSI at the transmitter (CSIT). In some typical applications of IoT, such as ambient or back-scatter communications, obtaining accurate CSIT of devices may be difficult. In this case, limited statistical or unavailable CSIT of devices is more likely to be present. How can we exploit the advantages of the CI with limited statistical or unavailable CSIT of the devices? The energy-efficient multi-cluster coordination in this case is not yet well understood.
4) **Joint device detection and channel estimation in grant-free multi-cluster coordination**: Low latency is an important metric for some IoT

applications, such as early warning of malfunctioned equipment. However, the complicated handshake mechanism that has been extensively used for grant access in cellular networks, leads to high access delay and outdated CSI. Considering the sporadic transmission at uplink, joint active device detection and channel estimation would benefit system performance in terms of low access delay and accurate CSI acquiring. Since in practice, the number of active devices may be much lower than the total number of devices, it is reasonable to apply compressed sensing theory for the device detection based on the transmission sparsity. The integration of NOMA and grant-free has been extensively researched, however, how to further improve the detection performance with the multi-cluster coordination needs more fundamental analysis.

5) **AI-aided non-orthogonal transmission in a multi-cluster coordination IoT**: In IoT scenario, the actions and operations of the devices may be programmed, and thus their positions, motions, and the associated channel quality could be observable, predictable, and learnable. Especially, the periodic transmission is incurred by the devices at uplink, such as the periodic monitoring and reporting of sensors, as well as the control feedback of robotic arms. These regularities lay a strong foundation for the AI-aided non-orthogonal transmission designs, in terms of cluster association, device grouping, massive device access, coordination level selection, and resource allocation, etc. However, the burst transmission of the devices harms the regularities in IoT, complicates the learning process, and impairs the testing performance. This again requires fundamental analysis and designs for the learning-aided non-orthogonal transmission in the multi-cluster coordination IoT.

8.5 Conclusion and Remarks

In this chapter, we introduced disruptive approaches for inter-cluster and intra-cluster interference management for multi-cluster coordinated IoT. Focusing on the typical MMTC scenarios in multi-cluster coordination, we have discussed the potential of accommodating non-orthogonal multi-device access, and even exploiting the ubiquitous interference as a green source by CI for enhancing reception performance. These novel solutions enable an energy-efficient transmission, and offer a new design philosophy of interference management for the emerging multi-cluster coordinated IoT. Challenges relating to emerging multi-cluster applications are also envisaged, and these give promise of exciting and insightful research over the years to come.

9

Conclusion of the Book

Internet of Things (IoT) provides enormous connections for massive devices and sensors, which has been considered as an enabling technology to build smart homes, smart transportations, smart cities, and smart planets. It is expected that by 2030 over 500 billion IoT devices will be in use with functions of sensing, computing, and communications, in which a surging amount of data and information exchanges are carried out among various IoT devices. Since IoT devices have sparse activity and low signaling overhead, random access (RA) is an efficient means to relieve the burden of machine-type communication (MTC). Considering that efficient signal detection methods are crucial to IoT RA systems, in this book, we focus on the signal detection for IoT, which is shown as follows.

In Chapter 3, in order to decide the activity variables of aggregated IoT signals, some low-complexity approaches, e.g., variational inference (VI) algorithm, are introduced. Since that the signal vector contains a number of zero elements, the information vector being transmitted becomes sparse. In this case, the compressive sensing (CS) reconstruction algorithms, e.g., block compressed sensing (BCS) and structured subspace pursuit (SSP) algorithm, can be applied to detect sparse signals.

However, when massive IoT devices access the network, the preamble collision occurs frequently and data packets from the device experiencing collision may not be successfully decoded at the receiver. Thus, in Chapter 4, some approaches are described to detect the collided signals, e.g., automatic modulation classification (AMC) approach.

Meanwhile, the technique of non-orthogonal random access (NORA) utilizes the different time of arrival to identify multiple devices with the identical preamble to improve the throughput, the detection of the round-trip delay (RTD) in NORA becomes significant. In Chapter 5, some estimation methods of RTD of multiple signals in NORA are presented, e.g., maximum likelihood (ML) algorithm, VI algorithm, and Markov Chain Monte Carlo (MCMC) algorithm, respectively.

IoT Signal Detection, First Edition. Rui Han, Jingjing Wang, Lin Bai, and Jianwei Liu.

Besides, since backscatter communications are beneficial for power-limited or battery-free IoT devices to transmit their data, where the signal detection at the receiver becomes crucial. In Chapter 6, signal detection approaches for backscatter devices are proposed to detect the presence of parasite devices and separate parasite signals from host signals, e.g., Central Limit Theorem (CLT) criterion.

Finally, wireless communications in IoT cause severe access congestion due to insufficient frequency, time, power, antennas, and other resources, evidently. In Chapter 7, we analyze the throughput and power consumption of non-orthogonal multiple access (NOMA) systems and investigate the energy efficiency maximization problem of a full-duplex cooperative NOMA system. Furthermore, one of the efficient techniques is to cluster massive devices into a number of small groups, with some resources being shared and reused in different clusters, referred to as a multicluster IoT.

Therefore, in Chapter 8, signal design approaches for multicluster coordination are proposed, e.g., successive interference cancellation (SIC) design and constructive interference signal design.

References

1 H. Zeng, J. Wang, Z. Wei, et al. Multicluster-coordination Industrial Internet of Things: The era of nonorthogonal transmission. *IEEE Vehicular Technology Magazine*, 17(3):84–93, 2022.

2 K. Ashton. That 'Internet of Things'. *RFID Journal*, 22:97–114, 2009.

3 L. Atzori, A. Iera, and G. Morabito. Understanding the Internet of Things: Definition, potentials, and societal role of a fast evolving paradigm. *Ad Hoc Networks*, 56:122–140, 2017.

4 G. Singh, J. Singh, D. Mitra, et al. A roadmap toward prospects for IoT enabled 5G networks. In *2023 7th International Conference on Computing Methodologies and Communication (ICCMC)*, pages 1405–1410, 2023.

5 China Academy of Information and Communications Technology.The white paper on the IoT (2011). *Public Security in China (Comprehensive Edition)*, 7(3):148–152, 2012.

6 K. Katsalis, N. Nikaein, E. Schiller, et al. Network slices toward 5G communications: Slicing the LTE network. *IEEE Communications Magazine*, 55(8):146–154, 2017.

7 J. Ordonez-Lucena, P. Ameigeiras, D. Lopez, et al. Network slicing for 5G with SDN/NFV: Concepts, architectures and challenges. *IEEE Communications Magazine*, 55(5):80–87, 2017.

8 L. Chettri and R. Bera. A comprehensive survey on Internet of Things (IoT) toward 5G wireless systems. *IEEE Internet of Things Journal*, 7(1):16–32, 2020.

9 M. Yang, S. Lim, S.-M. Oh, et al. An uplink transmission scheme for TSN service in 5G industrial IoT. In *2020 International Conference on Information and Communication Technology Convergence (ICTC)*, pages 902–904, 2020.

10 P. A. Hafeez, G. Singh, J. Singh, et al. IoT in agriculture and healthcare: Applications and challenges. In *2022 3rd International Conference on Smart Electronics and Communication (ICOSEC)*, pages 446–450, 2022.

IoT Signal Detection, First Edition. Rui Han, Jingjing Wang, Lin Bai, and Jianwei Liu.

11 S. Kar, P. Mishra, and K.-C. Wang. 5G-IoT architecture for next generation smart systems. In *2021 IEEE 4th 5G World Forum (5GWF)*, pages 241–246, 2021.

12 Y. P. E. Wang, X. Lin, A. Adhikary, et al. A primer on 3GPP narrowband Internet of Things (NB-IoT). *IEEE Communications Magazine*, 55(3):117–123, 2016.

13 R. Han, M. Liu, J. Wang, et al. Anti-jamming strategy for satellite Internet of Things: Beam switching and optimization. *IEEE Internet of Things Journal*, 10(23):20254–20263, 2023.

14 Y. Meng, W. Zhang, H. Zhu, et al. Securing consumer IoT in the smart home: Architecture, challenges, and countermeasures. *IEEE Wireless Communications*, 25(6):53–59, 2019.

15 D.-D. Tran, V. N. Ha, and S. Chatzinotas. Novel reinforcement learning based power control and subchannel selection mechanism for grant-free NOMA URLLC-enabled systems. In *2022 IEEE 95th Vehicular Technology Conference: (VTC2022-Spring)*, pages 1–5, 2022.

16 H. Xiao, B. Ai, and W. Chen. Grant-free massive machine-type communications with backward activity detection. In *2019 IEEE Global Communications Conference (GLOBECOM)*, pages 1–6, 2019.

17 L. Liu, E. G. Larsson, W. Yu, et al. Sparse signal processing for grant-free massive connectivity: A future paradigm for random access protocols in the Internet of Things. *IEEE Signal Processing Magazine*, 35(5):88–99, 2018.

18 Y. D. Beyene, R. Jäntti, and K. Ruttik. Random access scheme for sporadic users in 5G. *IEEE Transactions on Wireless Communications*, 16(3):1823–1833, 2017.

19 C. Bockelmann, N. K. Pratas, G. Wunder, et al. Towards massive connectivity support for scalable mMTC communications in 5G networks. *IEEE Access*, 6:28969–28992, 2018.

20 R. Ghods, C. Jeon, A. Maleki, et al. Optimal large-MIMO data detection with transmit impairments. In *2015 53rd Annual Allerton Conference on Communication, Control, and Computing (Allerton)*, pages 1211–1218, 2015.

21 E. Biglieri, G. Taricco, and A. Tulino. Performance of space-time codes for a large number of antennas. *IEEE Transactions on Information Theory*, 48(7):1794–1803, 2002.

22 J. Zheng and Y. Sun. Generalized space-time shift keying: Randomized bitwise Markov chain Monte Carlo detection and code design. In *2013 IEEE International Conference on Signal Processing, Communication and Computing (ICSPCC 2013)*, pages 1–4, 2013.

23 H. Zhu, B. Farhang-Boroujeny, and R.-R. Chen. On performance of sphere decoding and Markov chain Monte Carlo detection methods. *IEEE Signal Processing Letters*, 12(10):669–672, 2005.

24 W. Kim and K. M. Lee. Continuous Markov random field optimization using fusion move driven Markov chain Monte Carlo technique. In *2010 20th International Conference on Pattern Recognition (ICPR 2010)*, pages 1364–1367, 2010.

25 D. M. Blei, A. Kucukelbir, and J. D. McAuliffe. Variational inference: A review for statisticians. *Journal of the American Statistical Association*, 112(518):859–877, 2017.

26 C. M. Bishop and N. M. Nasrabadi. *Pattern Recognition and Machine Learning*, volume 4. Springer, 2006.

27 D. L. Donoho. Compressed sensing. *IEEE Transactions on Information Theory*, 52(4):1289–1306, 2006.

28 N. Serafimovski, A. Younis, R. Mesleh, et al. Practical implementation of spatial modulation. *IEEE Transactions on Vehicular Technology*, 62(9):4511–4523, 2013.

29 A. Garcia-Rodriguez and C. Masouros. Low-complexity compressive sensing detection for spatial modulation in large-scale multiple access channels. *IEEE Transactions on Communications*, 63(7):2565–2579, 2015.

30 R. Han, L. Bai, W. Zhang, et al. Variational inference based sparse signal detection for next generation multiple access. *IEEE Journal on Selected Areas in Communications*, 40(4):1114–1127, 2022.

31 D. Tse and P. Viswanath. *Fundamentals of Wireless Communication*. Cambridge University Press, 2005.

32 J. Choi. *Optimal Combining and Detection*. Cambridge University Press, 2010.

33 R. Y. Mesleh, H. Haas, S. Sinanovic, et al. Spatial modulation. *IEEE Transactions on Vehicular Technology*, 57(4):2228–2241, 2008.

34 A. Younis, N. Serafimovski, R. Mesleh, et al. Generalised spatial modulation. In *2010 Conference Record of the Forty Fourth Asilomar Conference on Signals, Systems and Computers*, pages 1498–1502, 2010.

35 L. Applebaum, W. U. Bajwa, M. F. Duarte, et al. Asynchronous code-division random access using convex optimization. *Physical Communication*, 5(2):129–147, 2012.

36 G. Wunder, P. Jung, and W. Chen. Compressive random access for post-LTE systems. In *2014 IEEE International Conference on Communications Workshops (ICC)*, pages 539–544, 2014.

37 J. Choi. Stability and throughput of random access with CS-based MUD for MTC. *IEEE Transactions on Vehicular Technology*, 67(3):2607–2616, 2018.

38 L. Xiao, P. Yang, Y. Xiao, et al. Efficient compressive sensing detectors for generalized spatial modulation systems. *IEEE Transactions on Vehicular Technology*, 66(2):1284–1298, 2017.

39 H. Zhu and G. B. Giannakis. Exploiting sparse user activity in multiuser detection. *IEEE Transactions on Communications*, 59(2):454–465, 2011.

40 Y. C. Eldar and G. Kutyniok. *Compressed Sensing: Theory and Applications*. Cambridge University Press, 2012.

41 M. Opper and C. Archambeau. The variational Gaussian approximation revisited. *Neural Computation*, 21(3):786–792, 2009.

42 L. L. Scharf. *Statistical Signal Processing: Detection, Estimation, and Time Series Analysis*. Addison-Wesley Publishing Company, 1991.

43 C. M. Biship. *Pattern Recognition and Machine Learning* (Information Science and Statistics). Springer, New York, 2007.

44 D. A. Harville. *Matrix Algebra from a Statistician's Perspective*. Taylor & Francis, 1998.

45 B. L. Sturm and M. Christensen. Comparison of orthogonal matching pursuit implementations. In *2012 Proceedings of the 20th European Signal Processing Conference (EUSIPCO)*, pages 220–224, 2012.

46 J. Choi, K. Lee, and N. Y. Yu. Compressive random access using multiple resource blocks for MTC. In *2016 IEEE Globecom Workshops (GC Wkshps)*, pages 1–5, 2016.

47 N. Jiang, Y. Deng, A. Nallanathan, et al. Analyzing random access collisions in massive IoT networks. *IEEE Transactions on Wireless Communications*, 17(10):6853–6870, 2018.

48 H. Shariatmadari, R. Ratasuk, S. Iraji, et al. Machine-type communications: Current status and future perspectives toward 5G systems. *IEEE Communications Magazine*, 53(9):10–17, 2015.

49 P. Patel and J. Holtzman. Analysis of a simple successive interference cancellation scheme in a DS/CDMA system. *IEEE Journal on Selected Areas in Communications*, 12(5):796–807, 1994.

50 L. Liu and W. Yu. Massive connectivity with massive MIMO—Part I: Device activity detection and channel estimation. *IEEE Transactions on Signal Processing*, 66(11):2933–2946, 2018.

51 J. G. Andrews, F. Baccelli, and R. K. Ganti. A tractable approach to coverage and rate in cellular networks. *IEEE Transactions on Communications*, 59(11):3122–3134, 2011.

52 T. Bai and R. W. Heath. Analyzing uplink SINR and rate in massive MIMO systems using stochastic geometry. *IEEE Transactions on Communications*, 64(11):4592–4606, 2016.

53 W. Xiao, R. Ratasuk, A. Ghosh, et al. Uplink power control, interference coordination and resource allocation for 3GPP E-UTRA. In *IEEE Vehicular Technology Conference*, pages 1–5, 2006.

54 M. Kiessling and J. Speidel. Analytical performance of MIMO zero-forcing receivers in correlated Rayleigh fading environments. In *2003 4th IEEE Workshop on Signal Processing Advances in Wireless Communications-SPAWC 2003*, pages 383–387, 2003.

55 J. Ding, D. Qu, H. Jiang, et al. Success probability of grant-free random access with massive MIMO. *IEEE Internet of Things Journal*, 6(1):506–516, 2018.

56 L. Bai and J. Choi. *Low Complexity MIMO Detection*. Springer, 2012.

57 L. Bai, J. Choi, and Q. Yu. *Low Complexity MIMO Receivers*. Springer, 2014.

58 E. Bjornson, E. De Carvalho, E. G. Larsson, et al. Random access protocol for massive MIMO: Strongest-user collision resolution (SUCR). In *2016 IEEE International Conference on Communications (ICC)*, pages 1–6, 2016.

59 J. Hoydis, S. Ten Brink, and M. Debbah. Massive MIMO in the UL/DL of cellular networks: How many antennas do we need? *IEEE Journal on Selected Areas in Communications*, 31(2):160–171, 2013.

60 Y. Han, G. Wei, C. Song, et al. Hierarchical digital modulation recognition based on higher-order cumulants. In *2012 2nd International Conference on Instrumentation, Measurement, Computer, Communication and Control*, pages 1645–1648, 2012.

61 Z. Zhang, X. Wang, Y. Zhang, et al. Grant-free rateless multiple access: A novel massive access scheme for Internet of Things. *IEEE Communications Letters*, 20(10):2019–2022, 2016.

62 F. Cong, A. K. Nandi, Z. He, et al. Fast and effective model order selection method to determine the number of sources in a linear transformation model. In *2012 Proceedings of the 20th European Signal Processing Conference (EUSIPCO)*, pages 1870–1874, 2012.

63 L. Lu, G. Y. Li, A. L. Swindlehurst, et al. An overview of massive MIMO: Benefits and challenges. *IEEE Journal of Selected Topics in Signal Processing*, 8(5):742–758, 2014.

64 R. Abbas, M. Shirvanimoghaddam, Y. Li, et al. A novel analytical framework for massive grant-free NOMA. *IEEE Transactions on Communications*, 67(3):2436–2449, 2018.

65 J. Liu, N. Kato, H. Ujikawa, et al. Device-to-device communication for mobile multimedia in emerging 5G networks. *ACM Transactions on Multimedia Computing, Communications, and Applications (TOMM)*, 12(5s):1–20, 2016.

66 J. Dai, J. Liu, Y. Shi, et al. Analytical modeling of resource allocation in D2D overlaying multihop multichannel uplink cellular networks. *IEEE Transactions on Vehicular Technology*, 66(8):6633–6644, 2017.

67 Y. Liang, X. Li, J. Zhang, et al. Non-orthogonal random access for 5G networks. *IEEE Transactions on Wireless Communications*, 16(7): 4817–4831, 2017.

68 C. Stefanovic, M. Momoda, and P. Popovski. Exploiting capture effect in frameless ALOHA for massive wireless random access. In *2014 IEEE Wireless Communications and Networking Conference (WCNC)*, pages 1762–1767, 2014.

69 G. Bartoli, N. C. Beaulieu, R. Fantacci, et al. An effective multiuser detection scheme for MPR random access networks. *IEEE Transactions on Communications*, 65(3):1119–1130, 2016.

70 J. Choi. NOMA-based random access with multichannel ALOHA. *IEEE Journal on Selected Areas in Communications*, 35(12):2736–2743, 2017.

71 R. Han, L. Bai, J. Yi, et al. Multiple delay estimation using genetic algorithm-based MCMC in non-orthogonal random access. *IEEE Wireless Communications Letters*, 9(3):398–401, 2020.

72 L. Bai, R. Han, J. Liu, et al. Multiple delay estimation for collision resolution in non-orthogonal random access. *IEEE Transactions on Vehicular Technology*, 69(1):497–508, 2020.

73 3GPP. Evolved universal terrestrial radio access (E-UTRA); base station (BS) radio transmission and reception. *Technical Specification 36.104, 3rd Generation Partnership Project (3GPP)*, October, 2014.

74 T. Morohashi, C.-H. Liao, A. Koizuka, et al. A high-performance RACH detection scheme for random access overload in LTE-advanced. In *2015 IEEE Conference on Standards for Communications and Networking (CSCN)*, pages 1–6, 2015.

75 M. I. Jordan, Z. Ghahramani, T. S. Jaakkola, et al. An introduction to variational methods for graphical models. *Machine Learning*, 37:183–233, 1999.

76 J. Choi. MCMC-based detection for random access with preambles in MTC. *IEEE Transactions on Communications*, 67(1):835–846, 2018.

77 L. Bai, T. Li, J. Liu, et al. Large-scale MIMO detection using MCMC approach with blockwise sampling. *IEEE Transactions on Communications*, 64(9):3697–3707, 2016.

78 G. Casella and E. I. George. Explaining the Gibbs sampler. *The American Statistician*, 46(3):167–174, 1992.

79 K.-S. Tang, K.-F. Man, S. Kwong, et al. Genetic algorithms and their applications. *IEEE Signal Processing Magazine*, 13(6):22–37, 1996.

80 Y. Sakamoto, M. Ishiguro, and G. Kitagawa. Akaike information criterion statistics. *Dordrecht, The Netherlands: D. Reidel*, 81(10.5555):26853, 1986.

81 A. A. Neath and J. E. Cavanaugh. The Bayesian information criterion: Background, derivation, and applications. *Wiley Interdisciplinary Reviews: Computational Statistics*, 4(2):199–203, 2012.

82 S. M. Kay. *Fundamentals of Statistical Processing: Detection Theory*, volume 2. Pearson Education India, 2009.

83 R. Han, L. Bai, Y. Wen, et al. UAV-aided backscatter communications: Performance analysis and trajectory optimization. *IEEE Journal on Selected Areas in Communications*, 39(10):3129–3143, 2021.

84 W. Zhang, Y. Qin, W. Zhao, et al. A green paradigm for Internet of Things: Ambient backscatter communications. *China Communications*, 16(7):109–119, 2019.

85 D. Dobkin. *The RF in RFID: UHF RFID in Practice*. Newnes, 2012.

86 R. Han, Y. Wen, L. Bai, et al. Rate splitting on mobile edge computing for UAV-aided IoT systems. *IEEE Transactions on Cognitive Communications and Networking*, 6(4):1193–1203, 2020.

87 L. Bai, R. Han, J. Liu, et al. Air-to-ground wireless links for high-speed UAVs. *IEEE Journal on Selected Areas in Communications*, 38(12):2918–2930, 2020.

88 Q. Zhang, H. Guo, Y.-C. Liang, et al. Constellation learning-based signal detection for ambient backscatter communication systems. *IEEE Journal on Selected Areas in Communications*, 37(2):452–463, 2018.

89 F. Rezaei, C. Tellambura, and S. Herath. Large-scale wireless-powered networks with backscatter communications—a comprehensive survey. *IEEE Open Journal of the Communications Society*, 1:1100–1130, 2020.

90 J. D. Griffin and G. D. Durgin. Gains for RF tags using multiple antennas. *IEEE Transactions on Antennas and Propagation*, 56(2):563–570, 2008.

91 A. Varshney, O. Harms, C. Pérez-Penichet, et al. LoRea: A backscatter architecture that achieves a long communication range. In *Proceedings of the 15th ACM Conference on Embedded Network Sensor Systems*, pages 1–14, 2017.

92 J. Wang, J. Zhang, R. Saha, et al. Pushing the range limits of commercial passive RFIDs. In *16th USENIX Symposium on Networked Systems Design and Implementation*, pages 301–316, 2019.

93 M. Y. Abdelsadek, Y. Gadallah, and M. H. Ahmed. Matching-based resource allocation for critical MTC in massive MIMO LTE networks. *IEEE Access*, 7:127141–127153, 2019.

94 K. Senel and E. G. Larsson. Grant-free massive MTC-enabled massive MIMO: A compressive sensing approach. *IEEE Transactions on Communications*, 66(12):6164–6175, 2018.

95 P. Popovski, C. Stefanovic, J. J. Nielsen, et al. Wireless access in ultra-reliable low-latency communication (URLLC). *IEEE Transactions on Communications*, 67(8):5783–5801, 2019.

96 E. Björnson, E. G. Larsson, and M. Debbah. Massive MIMO for maximal spectral efficiency: How many users and pilots should be allocated? *IEEE Transactions on Wireless Communications*, 15(2):1293–1308, 2015.

97 G. Yang, Y.-C. Liang, and Q. Zhang. Cooperative receiver for ambient backscatter communications with multiple antennas. In *2017 IEEE International Conference on Communications (ICC)*, pages 1–6, 2017.

98 W. Zhao, G. Wang, S. Atapattu, et al. Channel estimation for ambient backscatter communication systems with massive-antenna reader. *IEEE Transactions on Vehicular Technology*, 68(8):8254–8258, 2019.

99 H. Jiang, D. Qu, J. Ding, et al. Multiple preambles for high success rate of grant-free random access with massive MIMO. *IEEE Transactions on Wireless Communications*, 18(10):4779–4789, 2019.

100 H. Ding, D. B. Da Costa, and J. Ge. Outage analysis for cooperative ambient backscatter systems. *IEEE Wireless Communications Letters*, 9(5):601–605, 2019.

101 J.-B. Kim and I.-H. Lee. Capacity analysis of cooperative relaying systems using non-orthogonal multiple access. *IEEE Communications Letters*, 19(11):1949–1952, 2015.

102 Y. Liu, Z. Qin, M. Elkashlan, et al. Non-orthogonal multiple access for 5G and beyond. *Proceedings of the IEEE*, 105(12):2347–2381, 2017.

103 Z. Ding, M. Peng, and H. V. Poor. Cooperative non-orthogonal multiple access in 5G systems. *IEEE Communications Letters*, 19(8):1462–1465, 2015.

104 G. Liu, X. Chen, Z. Ding, et al. Hybrid half-duplex/full-duplex cooperative non-orthogonal multiple access with transmit power adaptation. *IEEE Transactions on Wireless Communications*, 17(1):506–519, 2017.

105 L. Zhang, J. Liu, M. Xiao, et al. Performance analysis and optimization in downlink NOMA systems with cooperative full-duplex relaying. *IEEE Journal on Selected Areas in Communications*, 35(10):2398–2412, 2017.

106 Z. Zhang, Z. Ma, M. Xiao, et al. Full-duplex device-to-device-aided cooperative nonorthogonal multiple access. *IEEE Transactions on Vehicular Technology*, 66(5):4467–4471, 2016.

107 J. Choi. Minimum power multicast beamforming with superposition coding for multiresolution broadcast and application to NOMA systems. *IEEE Transactions on Communications*, 63(3):791–800, 2015.

108 F. Fang, H. Zhang, J. Cheng, et al. Energy-efficient resource allocation for downlink non-orthogonal multiple access network. *IEEE Transactions on Communications*, 64(9):3722–3732, 2016.

109 Q. Sun, S. Han, I. Chin-Lin, et al. Energy efficiency optimization for fading MIMO non-orthogonal multiple access systems. In *2015 IEEE International Conference on Communications (ICC)*, pages 2668–2673, 2015.

110 Z. Zhang, K. Long, A. V. Vasilakos, et al. Full-duplex wireless communications: Challenges, solutions, and future research directions. *Proceedings of the IEEE*, 104(7):1369–1409, 2016.

111 B. Klaiqi, X. Chu, and J. Zhang. Energy-efficient and low signaling overhead cooperative relaying with proactive relay subset selection. *IEEE Transactions on Communications*, 64(3):1001–1015, 2016.

112 Y. Xiao and L. J. Cimini. Impact of overhead on spectral efficiency of cooperative relaying. *IEEE Transactions on Wireless Communications*, 12(5):2228–2239, 2013.

113 Z. Wei, X. Zhu, S. Sun, et al. Energy-efficiency of millimeter-wave full-duplex relaying systems: Challenges and solutions. *IEEE Access*, 4:4848–4860, 2016.

114 D. Nguyen, L.-N. Tran, P. Pirinen, et al. Precoding for full duplex multiuser MIMO systems: Spectral and energy efficiency maximization. *IEEE Transactions on Signal Processing*, 61(16):4038–4050, 2013.

115 B. Li, X. Qi, K. Huang, et al. Security-reliability tradeoff analysis for cooperative NOMA in cognitive radio networks. *IEEE Transactions on Communications*, 67(1):83–96, 2018.

116 Y. Tao, L. Liu, S. Liu, et al. A survey: Several technologies of non-orthogonal transmission for 5G. *China Communications*, 12(10):1–15, 2015.

117 T. Qiu, J. Chi, X. Zhou, et al. Edge computing in Industrial Internet of Things: Architecture, advances and challenges. *IEEE Communications Surveys and Tutorials*, 22(4):2462–2488, 2020.

118 K. Wang, Y. Wang, Y. Sun, et al. Green Industrial Internet of Things architecture: An energy-efficient perspective. *IEEE Communications Magazine*, 54(12):48–54, 2016.

119 J. An, K. Wang, S. Wang, et al. Antenna array calibration for IIoT oriented satellites: From orthogonal CDMA to NOMA. *IEEE Wireless Communications*, 27(6):28–36, 2020.

120 B. Selim, M. S. Alam, J. V. Evangelista, et al. NOMA-based IoT networks: Impulsive noise effects and mitigation. *IEEE Communications Magazine*, 58(11):69–75, 2020.

121 Q. Chen, K. Yang, H. Jiang, et al. Joint beamforming coordination and user selection for CoMP-enabled NR-U networks. *IEEE Internet of Things Journal*, 9(16):14530–14541, 2021.

122 P. Jia, X. Wang, and K. Zheng. Distributed clock synchronization based on intelligent clustering in local area industrial IoT systems. *IEEE Transactions on Industrial Informatics*, 16(6):3697–3707, 2019.

123 B. Cheng, M. Wang, S. Zhao, et al. Situation-aware dynamic service coordination in an IoT environment. *IEEE/ACM Transactions on Networking*, 25(4):2082–2095, 2017.

124 Y. Yuan, Z. Yuan, G. Yu, et al. Non-orthogonal transmission technology in LTE evolution. *IEEE Communications Magazine*, 54(7):68–74, 2016.

125 A. I. Perez-Neira, M. Caus, and M. A. Vazquez. Non-orthogonal transmission techniques for multibeam satellite systems. *IEEE Communications Magazine*, 57(12):58–63, 2019.

126 X. Liu and X. Zhang. NOMA-based resource allocation for cluster-based cognitive Industrial Internet of Things. *IEEE Transactions on Industrial Informatics*, 16(8):5379–5388, 2019.

127 Y. Cao, N. Zhao, Y. Chen, et al. Secure transmission via beamforming optimization for NOMA networks. *IEEE Wireless Communications*, 27(1):193–199, 2019.

128 M. S. Ali, E. Hossain, and D. I. Kim. Coordinated multipoint transmission in downlink multi-cell NOMA systems: Models and spectral efficiency performance. *IEEE Wireless Communications*, 25(2):24–31, 2018.

129 S. Khairy, P. Balaprakash, L. X. Cai, et al. Constrained deep reinforcement learning for energy sustainable multi-UAV based random access IoT networks with NOMA. *IEEE Journal on Selected Areas in Communications*, 39(4): 1101–1115, 2020.

130 Z. Wei, C. Masouros, K.-K. Wong, et al. Multi-cell interference exploitation: Enhancing the power efficiency in cell coordination. *IEEE Transactions on Wireless Communications*, 19(1):547–562, 2019.

131 H. Zeng, X. Zhu, Y. Jiang, et al. A green coordinated multi-cell NOMA system with fuzzy logic based multi-criterion user mode selection and resource allocation. *IEEE Journal of Selected Topics in Signal Processing*, 13(3):480–495, 2019.

132 Z. Wei, C. Masouros, F. Liu, et al. Energy-and cost-efficient physical layer security in the era of IoT: The role of interference. *IEEE Communications Magazine*, 58(4):81–87, 2020.

133 T. Xu, C. Masouros, and I. Darwazeh. Waveform and space precoding for next generation downlink narrowband IoT. *IEEE Internet of Things Journal*, 6(3):5097–5107, 2019.

Index

IoT Signal Detection, First Edition. Rui Han, Jingjing Wang, Lin Bai, and Jianwei Liu.

Printed and bound by CPI Group (UK) Ltd, Croydon, CR0 4YY
17/09/2026
14964308-0001